Case Studies in Plant Taxonomy

Exercises in Applied Pattern Recognition

Case Studies in Plant Taxonomy

Exercises in Applied Pattern Recognition

Tod F. Stuessy

Columbia University Press
New York

Columbia University Press
New York Chichester, West Sussex

Library of Congress Cataloging-in-Publication Data
Stuessy, Tod F.
Case studies in plant taxonomy: exercises in applied pattern recognition / Tod F. Stuessy.
p. cm.
Includes bibliographical references and index.
ISBN 0-231-07611-8
1. Botany—Classification. I. Title.
QK95.S77 1994
581′ .012—dc20 93-46062
CIP

Casebound editions of Columbia University Press books are printed on permanent and durable acid-free paper.

Printed in the United States of America
c 10 9 8 7 6 5 4 3 2 1

Contents

Preface

Most introductory courses in plant taxonomy concentrate on topics such as the historical development of systems of classification, nomenclature, family characteristics, local flora, and key writing. More advanced courses, variously entitled Experimental Taxonomy or Plant Biosystematics, attempt to build on this introduction by orienting around an evolutionary framework and by emphasizing how techniques such as cytology, chemistry, or statistics can improve our understanding of plant relationships. The texts for both types of courses, however, usually fail to provide exercises for students to apply what they learn. Students may come away from these courses with an understanding of what kinds of data are obtainable from plants, and perhaps how to go about gathering this information, but they lack the experience and confidence to make sound classificatory judgments.

Such a deficiency in students completing my own course in Experimental Taxonomy has been embarrassingly apparent (to the students as well as to me) on the final examinations. There, I usually include a hypothetical set of morphological, cytological, chemical, or other data and ask the students to determine their taxonomic significance. In general they perform poorly. They also have raised the question as to why this same type of exercise was not part of their course work.

This book attempts a remedy. It provides case studies taken from the botanical systematic literature. But these are not case studies in the usual sense; this book is not a compendium of reprints of papers on different topics (e.g., Ornduff 1967). Rather, I have drawn data from papers and summarized the discussion so as not to reveal the taxonomic relationships inferred by the author (see Stuessy 1977 for a report on the value and efficacy of this approach). The data are provided—but not the interpretations. That phase is left to the student.

Before plunging the student into case study exercises, several chapters present background, one chapter presents hypothetical cases with interpretive reasoning and explanation, and another chapter provides three case study examples, again with explanatory discussion. Ten case study exercises then follow. In each of these, the student is asked to classify groups in the context of a particular level in the taxonomic hierarchy (e.g., specific limits, varietal delimitation). The book thus aims to help the student become more adept in (1) recognizing similarities and differences in sets of comparative data; and (2) interpreting what these similarities and differences mean with reference to the various ranks in the Linnaean hierarchy.

The ten case studies have been selected using three criteria. First, all were published in journals in the United States with high circulation and that regularly contain plant taxonomic papers, namely, *American Journal of Botany, Brittonia, Evolution,* and *Systematic Botany*. This should enable teachers in universities and colleges with limited periodical holdings to obtain the original articles with minimal effort. Second, the papers from which all the case studies were drawn displayed their data in tables and/or figures in such a manner that the graphics alone provide sufficient information for taxonomic decisions to be made. Interesting papers with lengthy discussions and limited visual display of data, therefore, were deemed unsuitable. Third, the cases chosen reflect different types of taxonomic problems as well as different types of data.

It is impossible within the confines of only ten case studies to display every type of data used and every type of problem encountered in taxonomic studies. Hence, I have selected studies in which morphology is emphasized along with other more experimental information, including cytology, cytogenetics, palynology, isozymes, and cpDNA. Different styles of graphics have also been included deliberately. The specific and infraspecific levels in the hierarchy have also been emphasized, because it is here that many of the fascinating problems occur in routine taxonomic work. The main objective is for the student to gain experience in making classifications—not in understanding every nuance of all types of comparative data, nor in all taxonomic contexts.

Inclusion of a particular paper in this book does not necessarily mean that I (or the students) agree with the interpretations made by the original author.

In any event, those original interpretations are not included here; the goal is for students to make their own interpretations. Remember, too, that the data in the case studies have been taken out of context for pedagogical reasons. The case study exercises do not reflect the full presentation of information or interpretation by the original author(s).

Another point needs to be stressed regarding the types of data selected for inclusion as exercises. In the past two decades data analysis in plant taxonomy (as well as in many other areas of biology) has become more quantitative. With the aid of computers, large amounts of data can now be gathered, quantitatively delimited, and statistically evaluated using any number of phenetic, cladistic, or phyletic algorithms (for a review of these approaches, see Stuessy 1990). The intuitive search for initial patterns of relationship, therefore, passes from the investigator to the computational machine. This new computational ability has advanced the field in many ways because computers can search for taxonomic structure within data sets in ways beyond the reach of the human mind. Certain kinds of data are generated in such quantities (e.g., from cpDNA) that there is almost no effective way to intuitively evaluate them; statistical evaluations are essential.

Nevertheless, in taxonomy it is still (and no doubt always will be) beneficial to develop one's own pattern recognition skills. At minimum, this ability can help the practitioner notice when computational results appear to be grossly wrong. This book, therefore, stresses the development of skills of *intuitive pattern recognition* and of the judgment necessary to apply those skills in solving botanical taxonomic problems.

The data chosen are specifically amenable to this type of analysis. In some cases the data presented might also profitably be compared via phenetic methods (e.g., in the sixth case study exercise), but this is not the intent of the exercise nor of the book itself. Likewise, chloroplast DNA data, now so very important in helping reconstruct phylogenies of plants (e.g., Palmer 1987; Crawford 1990; Sanderson and Doyle 1993), can be effectively used only by phenetic methods (for sequence data) and cladistic methods (for restriction site mutation data). As a result, this kind of information has been deemphasized in this book. One very simple (and early) cpDNA study is, however, included (the eighth case study exercise), but without the accompanying cladistic analysis. Likewise, one case study exercise (the seventh) deals with isozymes, but without the phenetic analysis contained in the paper.

Instructors who so wish will be able to easily extend these case studies into exercises in phenetic or cladistic analysis. These alternative approaches should be particularly meaningful after the students have struggled intuitively with the basic data. It is my firm belief, moreover, that without a clear intuitive grasp of relationships in a group of organisms, refined statistical analyses cannot be interpreted successfully. In fact, the new data and relationships may only be meaningful if used as a test of previous hypotheses generated intuitively by the worker who is well acquainted with the group in question. Because no type of data and no method of analysis provides absolute answers—largely because we lack a mathematical theory of clustering and because no phylogeny is known with certainty—this is the only reasonable way to proceed. Intuitive pattern recognition for development of initial hypotheses of relationships is, therefore, still vitally important in modern plant taxonomy and will likely remain so. This is the reason for the focus adopted here.

My experience in using the case studies might suggest the following classroom procedures: (1) assign a particular case study exercise to be worked through by each student individually; (2) then have each student put his or her classification on the blackboard (or overhead transparency or handout) with a brief presentation; (3) after all the classifications have been displayed, discuss the variation in results; (4) the instructor might then present the original author's classification and read any pertinent parts of the original paper; (5) follow with a final class discussion; and (6) put the original paper on library reserve so that students can peruse it at their convenience.

One of my most rewarding experiences as a teacher occurred while using the case studies. The class was Experimental Taxonomy and consisted of five graduate students, most of whom were in plant systematics and all of whom had strong botanical backgrounds. We were discussing the last case study in this book, which has more data than any of the others. The students presented their classifications on the blackboard, one by one. This particular case study is sufficiently complex that it generated a rich variation of student interpretations. The first four students presented their viewpoints, and discussion ensued on each. All of the classifications had good aspects, but no consensus was being reached. The last student then presented her own classification, and it was immediately obvious to the rest of the class that this was the correct approach. The other students volunteered comments such as "Oh, yes, that is much better than mine" and "Yes, yes, that feels much better; it seems to pull everything together better." It was rewarding to watch these dynamics in which one classification was accepted by the group as superior. This preferred classification, by the way, was nearly identical with that of the publishing author.

The background needed by students to profit from using this book will vary depending on the depth of analysis required by the professor. Obviously, students must have had some previous introduction to different types of comparative data ordinarily used for taxonomic purposes. Likewise, they should have had at least some exposure to the different ranks of the Linnaean hierarchy and how one evaluates these in light of different types of data. Furthermore, it would be helpful to have some acquaintance with processes of evolution in flowering plants. The ideal preparation might be the detailed study of two books, one dealing with taxonomic grouping and ranking (e.g., Stuessy 1990) and another dealing with plant evolution (e.g., Grant 1981).

Few students will have all this background. I have, therefore, attempted to fill the gaps somewhat by offering introductory chapters that sketch perspectives for the beginning student. In my experience *any* student will learn *a great deal* about decision making in taxonomy from any and all of these case studies (Stuessy 1977). Also, the case study exercises are ordered so that the more complex and difficult cases are encountered toward the end.

The case studies are useful for undergraduates as well as for graduate students. For lower-level students who might find some of the terms difficult, the instructor might pair this text with a standard text on plant systematics that contains a glossary.

Some of my colleagues might argue that the acquisition of discerning taxonomic judgment is a long and tedious process requiring many years of experience and that no set of ten exercises can adequately substitute. No doubt this is so, but in my opinion, much of the training traditionally acquired by years of trial-and-error experience can be more efficiently obtained by explicit coaching at the outset. Furthermore, these exercises reveal to the student how complex, challenging, and stimulating it is to solve taxonomic problems. That is, after all, one of the reasons we ourselves find the field so attractive.

Acknowledgments

I have received help in the preparation of this book from many sources. Foremost have been the students who in the first place stimulated me to use case studies in teaching concepts of plant classification. Feedback from students has also been invaluable in helping me choose the particular exercises used in this book—which combines up-to-date examples with more classical works.

Heartfelt appreciation goes to John H. Beaman, Samuel B. Jones, Jr., David J. Keil, and Carol R. Sweeney for helpful comments on an earlier draft of the manuscript. Dave Dennis redrew several figures from original sources. I also thank the following authors and journals (with copywriting institution) for generously granting permission to use their publications as case study examples or exercises: C. R. Annable, J. R. Ballington, D. Coates, W. J. Elisens, J. R. Estes, L. D. Gottlieb, J. W. Hardin, S. B. Jones, Jr., W. B. Kirkman, W. M. Klein, J. W. McClure, C. R. Parks, W. W. Payne, K. A. Petersen, K. M. Peterson, P. M. Peterson, A. M. Solomon, W. P. Stoutamire, *American Journal of Botany* (Botanical Society of America), *Brittonia* (New York Botanical Garden), *Evolution* (Society for the Study of Evolution), and *Systematic Botany* (American Society of Plant Taxonomists).

Finally, but very importantly, have been the editing and production staff of Columbia University Press, especially Anne McCoy and Laura Wood, who clarified and refined numerous of my fuzzy ideas regarding the manuscript. Heartfelt thanks also go to Connie Barlow for really outstanding copy editing.

Case Studies in Plant Taxonomy

Exercises in Applied Pattern Recognition

1. Taxonomy and Classification

Taxonomy is one of the most challenging areas of the biological sciences. Nevertheless, to some people it may bring forth images of dusty libraries, old museum collections, and stoop-shouldered, bespectacled curators. The truth is that the working conditions and people involved with taxonomy are varied. For example, teams of scientists often work on collaborative projects with sophisticated field and laboratory equipment, such as with electron microscopes and high-speed computers. Taxonomists do classify and identify the world's biota but they also work on exciting conceptual and methodological questions that relate directly to other sciences as well as to societal issues. To better understand what taxonomy is, let us turn first to some related terms and definitions.

Definitions

Taxonomy has had various meanings since it was first coined by Augustin Pyramus de Candolle in 1813, and particular confusion has prevailed with *systematics*. Before the theory of evolution by means of natural selection was proposed by Charles Darwin, the two words were regarded as more or less synonymous. Prior to 1859 botanists and zoologists interested in relationships studied and compared morphological features in search of patterns for grouping organisms into units. This ordering of organisms into groups or *taxa* (sing. *taxon*) based on similarities and/or differences was (and still is) called *classification*. In pre-Darwinian times a widely accepted classification was often believed to reflect the plan of natural order created by God, and humans were simply attempting to reveal this Divine Plan. Naturalists engaged in these activities of classification were called interchangeably taxonomists or systematists.

Since Darwin's time, not only have systematists continued their interest in classification but they have also attempted to understand the evolutionary relationships among the groups so ordered. Furthermore, some systematists are now primarily interested in providing a better understanding of the process of evolution itself, that is, of the mechanisms that produce the diversity of life. Consequently, a systematist today may study many different aspects of evolutionary biology that are far removed from the morphological investigations of a century ago. One might, for example, analyze the patterns of cytological variation within a particular taxon, or compare sequences of base pairs in chloroplast DNA molecules, or investigate the genetic basis of morphological variation within populations.

Because these and other studies spawned by evolutionary theory began to cover a wide range, a collective term was needed to designate them all, and another term was needed for describing the more traditional activities of classification. As a result, the term systematics has come to have a meaning different from and broader than taxonomy. The definition used by most people today is expressed well by Simpson (1961:7), who defines systematics as "the study of the kinds and diversity of organisms and any and all relationships among them." Some workers still equate taxonomy with systematics, but taxonomy is now more properly limited to the study of the principles and methods of classification (Simpson 1961). The relationship of these three terms is shown in figure 1.1; table 1.1 summarizes the goals and methodologies used in systematics.

Two other terms, *nomenclature* and *identification,* are sometimes confused with classification and systematics. After organisms have been classified, names must be given to these groups in accordance with rules of nomenclature. Identification, on the other hand, is referring an individual specimen to a previously classified and named group. For example, if one picks a small branch with leaves from a tree in the neighborhood, what is being sought is identification of the specimen—not a classification. Many years ago some taxonomist did classify the species represented by that individual tree, but the act of classification involved a great deal more effort—and usually more complete samples, especially of the tree's flowers and/or fruits.

Identification and nomenclature, therefore, are directly associated with

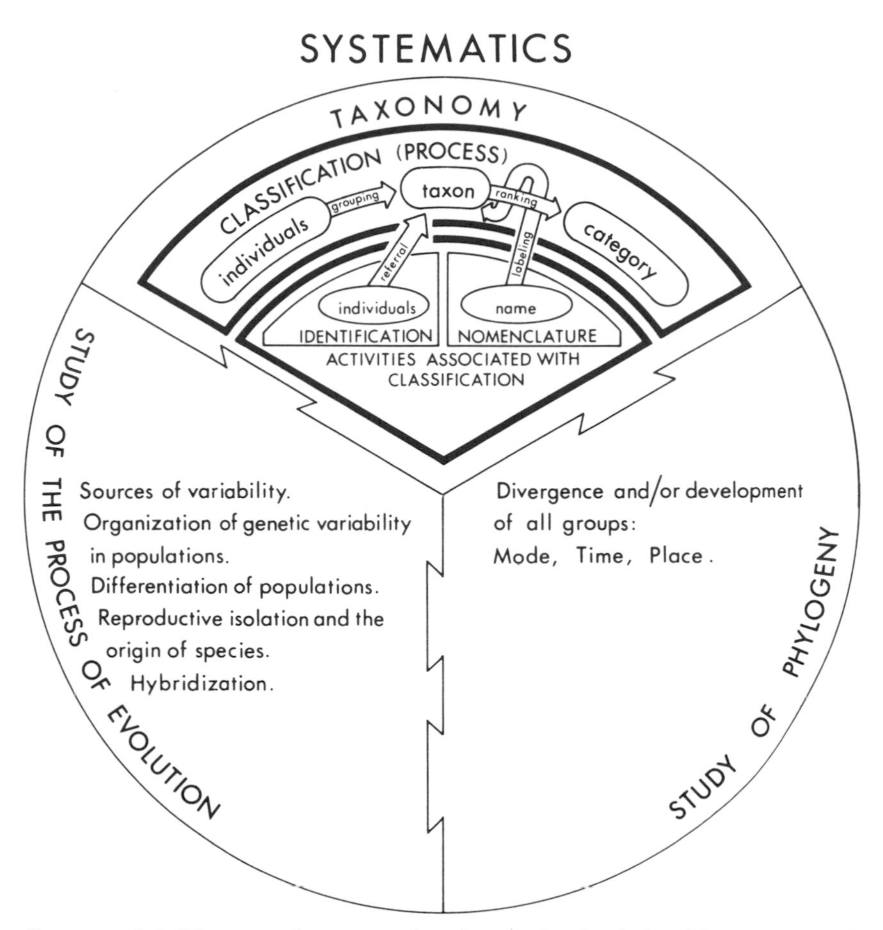

FIGURE 1.1 Diagram of conceptual and procedural relationships among and within areas of systematics. The tighter connection between the bottom two areas emphasizes their closeness as aspects of evolutionary processes—short term on the left, and long term on the right (from Stuessy 1979:622).

classification (figure 1.1), but one first classifies, then names, and then identifies subsequent samples. All three activities are part of the discipline of taxonomy. Taxonomy, in turn, is one aspect of systematics.

The Importance of Systematics in Biology

For purposes of providing a broad perspective of the significance of taxonomy, a discussion of the wider role that systematics plays in biology is in

TABLE 1.1.
Outline of Goals and Methodologies of Systematics

I. Accumulation of Comparative Data
   A. From the Organism
      1. Structures
      2. Processes (interactions among structures)
   B. From the Organism-Environment Interactions
      1. Distribution
      2. Ecology
II. Use of Comparative Data to Answer Specific Questions
   A. Classification (most predictive system of classification at all levels)
      1. Method and result of grouping of individuals
      2. Level in the taxonomic hierarchy at which the groups should be ranked
   B. Process of Evolution
      1. Nature and origin of individual variation
      2. Organization of genetic variation within populations
      3. Differentiation of populations
      4. Nature of reproductive isolation and modes of speciation
      5. Hybridization
   C. Phylogeny (divergence and/or development of all groups)
      1. Mode
      2. Time
      3. Place

SOURCE: Stuessy 1990:7

order (after Mayr 1968). A unifying goal of systematics is understanding the evolution of life on earth. Achievements in systematics not only help explain how extant and extinct organisms are related to each other but also provide a framework within which systematists can interpret the precise mechanisms of evolution. Such knowledge, in turn, is useful in other areas of biology, including population genetics, ecology, and cytogenetics.

Because of its broad and synthetic nature, systematics not only draws on information from other areas of biology but also acts as stimulus to these same areas. It is not hard to imagine how data about plant relationships might be useful to a population geneticist, a cytogeneticist, a physiologist, a morphologist, or a biogeographer. The comparative data generated in systematic studies answer or suggest questions in almost every other field of biology.

An obviously important role of systematics in biology is that of its component field and the subject of this book: taxonomy. By providing a system of classification and names for all life on earth, this aspect of systematics is essential for communication about any and all aspects of organisms. The classification serves as an organized outline or framework from which information about particular organisms or taxa can be logged and then retrieved and applied by others. These and other related aspects of classification are covered next.

The Importance and Universality of Classification

The term *classification* has a dual meaning in taxonomy, as it does colloquially. It refers to both a process and a product. As process it is the act of grouping and ranking organisms based on criteria of relationship. As product a classification is the resultant hierarchy of taxa. This latter usage sometimes also includes the characteristics used to group the ranked units. The importance of classification to biology and society results from both the process and the product of classification.

It has been quipped by Hopwood (1959:230) that the urge to classify is "like the predisposition to sin; it accompanies us into the world at birth and stays with us to the end." Although the quotation is somewhat tongue-in-cheek, it nevertheless expresses an important aspect of human nature. Classification of any set of objects, whether animate or inanimate, is basic to and essential for human society. If we reflect for a moment on the inanimate objects in our environment, we soon realize that we have classifications for nearly everything. Tables come in all shapes and sizes, and these may be further classified into dining, kitchen, picnic, coffee, card, end, pool, and so on. Furniture in general is arranged into periods, and even within each period variation in craftsmanship is recognized and appropriately labeled. Consider, too, the multiple or hierarchical categories by which your car is classified on your certificate of vehicle registration.

Classification also allows for the organization of information about objects in such a way that the resultant hierarchy of classes and associated names serves as a reference system of information for these items. For example, the fact that a Ford and a Chevy both belong within the larger class "automobile" tells us a lot about the general features of each: how they are constructed and how they should look both inside and out. Similarly, the position that groups of organisms have in a classification tells us what features they have in common.

A classification system is a prerequisite for both nomenclature and identification. The system of classification serves as a guide for labeling unnamed objects by comparison with those items already named. Objects are never absolutely identical to each other—not even machine-made pieces are identical—but they can be highly similar. The degree of similarity needed for two objects to be regarded as effectively identical, and thus given the same name, depends upon how detailed is the classification—the more detailed or dissected the hierarchy, the more similar the objects must be to carry the same name.

In addition to the functions already mentioned, most biological classifications also attempt to reflect evolutionary relationships. The degree to which biological classifications are held to be successful in doing this varies with the specific group and the level of taxon involved, and it must be admitted that in most cases it simply is not possible to know if the relationships expressed actually do reflect the ancestry of a group. However, if we assume that organisms have evolved, then we would expect that a comparative study of features of organisms should theoretically give us a good idea of how they are related evolutionarily. In any case, most biological classifications attempt to show these kinds of relationships; insight into evolutionary history is therefore one important contribution of both the process and product of classification. Classifications can also be regarded as scientific hypotheses of relationships; they can be tested by new data as more information becomes available for the organisms involved.

Overall, and from a general perspective, classifications are important because they are useful as summarizing and predictive devices (Warburton 1967). They hold together information about objects in an orderly fashion, and they also enable retrieval of this information. More importantly, however, a classification allows for predictions to be made—that is, taxa are constructed about which we can make inductive generalizations. For example, within taxon A, defined by features other than leaf arrangement, members of subtaxa X and Y are discovered to have opposite leaves. We can infer, therefore, that members of a third subtaxon Z in the same class should also have opposite leaves. To test this prediction in an informal sense, we can look at Z and see what type of leaf arrangement it has. To test the prediction even further, we can look at the leaves of the most distantly related subtaxon that is still included within taxon A. The ability to make such predictions is the basis for much of our search for useful plant materials, especially for chemical compounds that have medicinal value. If a useful compound is discovered in one group of plants, the most closely related groups are looked at next to learn if they, too, might have the same or similar compounds, perhaps with more desirable pharmacological value.

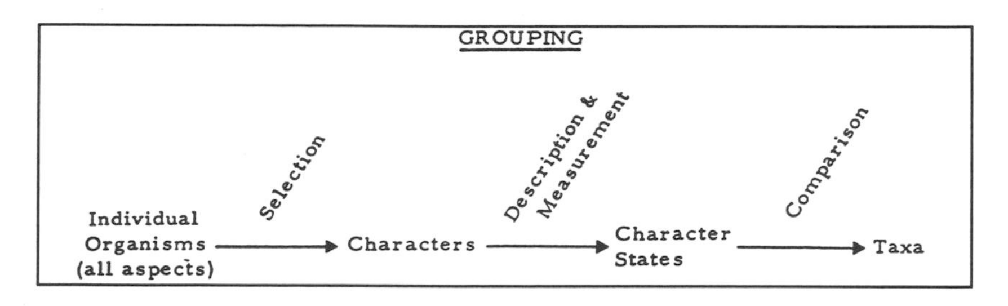

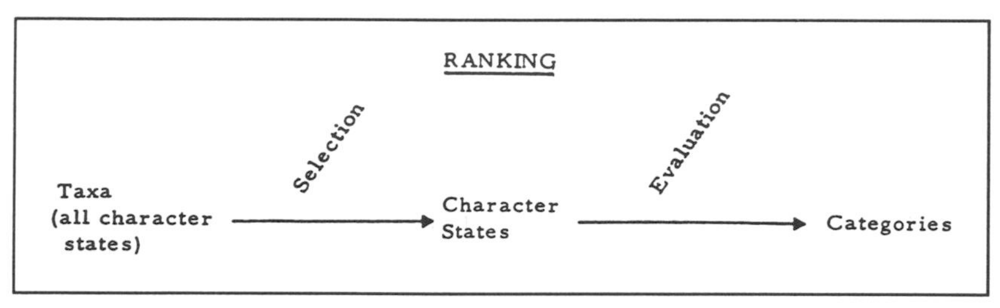

FIGURE 1.2 Representation of the two primary operations of classification (from Stuessy 1979:623).

The Anatomy of Classification

Earlier in this chapter, classification was defined as the ordering of organisms into groups based on observed similarities and/or differences. When only two or very few groups result from this process, we can treat them as being at the same level. But usually many units are involved, in which case we need to show the relationships *among* the groups for more effective communication about them. If many units have been created, we face the same problem as if we were confronting many separate individuals. To solve this difficulty, larger groups composed of smaller units are made and these are given different categorical names. The result is a taxonomic hierarchy of ranked units, with the largest units composed of smaller units; these, in turn, are built from still smaller units, and so on.

The process of classification, therefore, usually involves two separate operations: (1) grouping and (2) ranking. These two aspects are diagrammed in figure 1.2. Grouping involves three specific operations. First, one must select characteristics of the organisms to use in assessing similarities and differences. It is impossible to compare two or more objects without referring to specific features of each, and these selected aspects are called *taxonomic characters*. All taxonomic characters are also characteristics, but not all characteristics of organisms are regarded as taxonomic characters, especially those that are too variable or show no meaningful correlations with other features. The additional terms *feature* and *aspect* are treated as being synonymous with characteristic.

After selecting particular features to be treated as taxonomic characters, the second operation in grouping involves describing and/or measuring those characters. One cannot use the character "leaf shape," for example, to compare two plants meaningfully, because both plants have leaves with shape. Instead, the particular leaf shapes in the two plants must be compared, such as "obovate" versus "lanceolate." These latter are called *character states,* which are defined as any expression of a character that can be described (and/or named) and measured, and then used for purposes of taxonomic comparison and evaluation.

Character states can be either qualitative or quantitative. Qualitative characters are those that have character states given in descriptive terms. Examples of such characters include colors of petals and sculpturing of seed coats. Quantitative characters, on the other hand, are those that have character states in measurable units, such as number of petals in a flower, or size of leaves (length and/or width). In general, number and size of parts of a plant are regarded as quantitative aspects of structure, and these two aspects are assessed by counting and by measurement of linear dimensions, respectively. Shape, however, is a more complex feature, and this can be assessed qualitatively, quantitatively or both. For example, a narrow-leaf willow tree has lanceolate leaves, with dimensions of about 12 cm by 2 cm.

In addition to qualitative and quantitative, taxonomic characters can be distinguished as "good" and "bad." A good character is one that serves adequately to differentiate taxonomic groups, either in a taxonomic key or in some other way. A bad character is one that cannot be relied upon for such discrimination. Good characters are usually constant or vary only within defined limits within a taxonomic group and show correlations with other features, but bad characters vary unpredictably within groups, are often environmentally plastic, and show little or no correlation with other features. A bad character in one taxon (e.g., within a genus) may, however, turn out to be a good character in another.

Once characters have been selected and character states described and measured, the third operation in grouping is to compare the character states to obtain the groups. These comparisons can be made in different ways, and there still are important areas of disagreement regarding taxonomic philosophy and methods. Some workers emphasize overall similarity of states (phenetics), others stress only features they believe have come directly from evolutionary ancestors (cladistics), and still others rely on conditions of direct ancestors plus evolutionary changes within lineages (phyletics). Whatever the perspective, it is customary that some character states will be regarded as being more important, and these will carry more weight in the grouping decisions.

The next procedure in classification is ranking of the recognized groups. Ranking involves two specific operations. First, all the character states of the groups are examined, and some are selected for use. The character states might be the same and even include all of those used for grouping, but usually not all are used for ranking. Other characteristics also might be selected for consideration that were not used for grouping. Second, these selected character states are evaluated in terms of categories available for use in the taxonomic hierarchy. The presence or absence of certain kinds of features usually will suggest an appropriate rank in the hierarchy of classification.

2. The Taxonomic Hierarchy

In the previous chapter, the discussions of classification centered primarily on grouping. In this chapter, the emphasis moves to the task of ranking. Ranking is extremely important; without this second part of the process of classification, few benefits can be realized. Ranking has at least two essential functions: (1) it allows for a more efficient and consistent method of communication about the taxonomic units; and (2) it allows for the taxonomic units to be placed in categories that reflect degrees of differences in characters and character states, with those differences at the higher levels of ranking being very great, and those at lower levels being small.

The Taxonomic Hierarchy

In order for ranking to be achieved, a hierarchy of categories must be provided into which taxonomic units can be placed. For organisms, such a structure is called the Linnaean hierarchy. The Swedish botanist Carl Linnaeus used this hierarchy consistently in the mid-eighteenth century (e.g., Linnaeus 1753).

With inanimate objects, many different types of hierarchies, or sets of classes, can be constructed and used. With organisms, however, only one set of classes is acceptable for use, and this rigidity prevails for at least three reasons. First, for purposes of efficient and exact communication on a worldwide basis, one standard hierarchy is essential. Second, one particular category, the species, is fundamental to our understanding of organic diversity, and therefore all other categories in the hierarchy must relate directly or indirectly to this level. Third, it is assumed that all differences in the tree of life originated in the same general way—through evolution by natural selection—and therefore the units used to classify this diversity should apply equally well to all or any part of the living world.

Actually, the Linnaean hierarchy was developed decades before the idea of evolution was even entertained and more than a century before Darwin published *The Origin of Species* (1859). In the fifteenth and sixteenth centuries, a strong belief prevailed in the Divine Creation of organisms. This view of the origin and diversification of life did, however, recognize a pervasive similarity of organization throughout the living world. The early taxonomists believed their classifications were simply reflections of the Grand Plan of the Creator. Because different workers had different ideas of what God's grand design was really all about, it took several centuries for a taxonomic hierarchy of organisms to be developed and used consistently.

One person in particular, Linnaeus, helped stabilize these attempts to achieve a taxonomic hierarchy of organisms. Because of his influence in botany at that time and the simultaneous and almost universal recognition of the need to standardize systems of classification, his set of classes became the standard. The Linnaean hierarchy has since been modified slightly; the basic categories or units of the Linnaean hierarchy in use today by plant taxonomists are as follows:

Division
 Class
  Order
   Family
    Genus
     Species
      Subspecies
       Variety
        Form

Many additional intermediate categories in the hierarchy exist, such as subfamily, tribe, subgenus, section (see *International Code of Botanical Nomenclature;* Greuter et al. 1988), but the ones listed here are basic units.

Two general terms need to be defined and clarified: category and taxon. *Category* refers to a particular level (or rank) in the taxonomic hierarchy, such as genus or class; taken collectively, categories represent all the different levels of subdivisions available in the classification system. All categories may not always be used, of course, but they are available if needed. The second term, *taxon* (plural, taxa), refers to a group with shared features at any rank. Initial taxa may be grouped again to form taxa of higher rank. Category, thus, refers to an abstract level of classification; taxon is a particular group of organisms within a category. For example, our own species is *Homo sapiens; Homo* is the particular taxon within the category of genus; *Homo sapiens* is the particular taxon within the category of species.

The Linnaean hierarchy can be viewed as a system of classes within classes (or boxes within boxes), called "nested classes" (Buck and Hull 1966), or often referred to as a nested hierarchy. The adoption of a nested system did not come about fortuitously. One reason for its successful construction and employment was that it built upon less elaborate hierarchies in use at the time, rather than departing radically from schemes already in place. These earlier systems of classification, in turn, were also based on hierarchical concepts of relationship developed by the ancient Greeks, and in particular by Aristotle. His principle of "logical division" maintained that any group of objects could be divided into subgroups based upon a single criterion called a *fundamentum divisionis.* For example, all polygons can be divided into subgroups based on the *fundamentum divisionis* of number of angles—with triangles, squares, pentagons, hexagons, etc., resulting. The larger unit Aristotle called the "genus" and the smaller units the "species," terms which are still used in traditional formal logic (e.g., Sinclair 1951). This method of classification and the limited hierarchy of objects that resulted was a powerful precedent for a way to look at the world in an organized fashion. The Linnaean hierarchy, in fact, is simply an extension of the hierarchy developed by Aristotle in his concepts of logical division.

A second reason for the long-term success of the Linnaean hierarchy, however, must be credited to the genius of its inventor. Linnaeus understood the utility of a reasonable and fixed number of categories, and he ably demonstrated this utility by successfully classifying thousands of plants (and animals) known to him at that time (e.g., Linnaeus 1735, 1753). But a third and more fundamental reason exists for the efficacy of the Linnaean hierarchy. People everywhere and at all levels of cultural development classify animate and inanimate objects in their environment (e.g., Berlin 1992). The human brain cannot assess relationships (or connections) without assigning subordinate roles to objects and ideas. This intellectual ability must, to some extent, mirror the actual relationships of animate and inanimate objects, because our own cranial powers have also evolved on earth.

No absolute means currently exist for defining the basic units of the

Linnaean hierarchy. Not all botanical taxonomists, therefore, would define the categories in the same way. In fact, if asked to give criteria to circumscribe each of the commonly employed units, ten different systematists probably would give ten slightly different answers. Despite these variations, a ground of similarity prevails, and it is this similarity that is emphasized in the description of categories that follows.

Species

The species is the basic category in the taxonomic hierarchy. It is neither the highest-ranking category nor the lowest, but it is considered fundamental in biological classification for several reasons: (1) the species is believed to be the basic product of the evolutionary process; (2) the species is the smallest unit in the hierarchy that is usually recognized by all peoples of the world, whether culturally progressive or aboriginal and whether a trained botanist or a lay person; and (3) the species can be defined more explicitly and successfully by taxonomists than can any other category.

Although many workers have contributed to the development of the species concept, none has been so influential as the ornithologist Ernst Mayr. He early stressed that "the noninterbreeding of natural populations rather than the sterility of individuals be taken as the decisive species criterion" (1963:15; see also Mayr 1940). The definition of a species that results from this populational viewpoint is: "groups of actually or potentially interbreeding natural populations which are reproductively isolated from other such groups" (Mayr 1963:19). This is called *the biological species concept* and it is the species definition held today by most systematists.

Problems arise with the biological species concept, however, in that it is often difficult to determine whether or not two or more populations are actually interbreeding in nature. If individuals from two populations can be made to mate and successfully produce offspring in the laboratory, how does this pertain to interbreeding in the wild? The biological species concept, moreover, has a zoological bias. Had Ernst Mayr been a botanist, he might have promoted a different way to distinguish species because hybridization between otherwise quite morphologically distinct "species" is a common phenomenon in plants (e.g., Grant 1981).

Because of the difficulties encountered in applying the biological species concept, some biologists believe that a species is only a mental construct without objective existence (Burma 1954). Ehrlich and Raven (1969), Sokal and Crovello (1970), and Levin (1979) reason that the only truly valid evolutionary unit and the one most involved with speciation (Levin 1993) is the geographically isolated, interbreeding population. Although admitting that the category of species is needed for communication and discussion, they believe that any viewpoint or discussion that reifies the species concept—that treats species as something objectively real—should be rejected.

Other species concepts have also been introduced in recent years (for discussions see Stuessy 1989 and Ehreshefsky 1992). The one receiving most attention is the "phylogenetic" (cladistic) concept, which stresses that species are equivalent to terminal branches of cladograms (e.g., Donoghue 1985). The emphasis is on species as historical entities rather than as dynamic populational systems. This seems a return to sterile Aristotelian-like logic or Platonic essentialism, neither of which is recommended for defining the dynamics of populational divergence upon which species are based.

Although the application of the biological species concept is sometimes difficult in particular instances, usually it is workable—even in plant groups. By a careful evaluation of morphological, chemical, cytological, and other data, as well as of breeding relationships, identifiable populations usually can be assigned to one or more species without much difficulty. (For a useful practical approach, see Doyen and Slobodchikoff 1974). And although the category of species may not be defined exactly, it is at least as well defined and delimited as any of the other categories now in use in biology. "The species as a unit of organization is probably no more and no less universal and well defined than the individual, cell, gene, atom, or any other unit with which we have to deal" (Grant 1963:342).

Subspecies, Variety, and Form

The categories of subspecies, variety, and form have had changing definitions over the past two hundred years. Because of these changes in concept, and because the categories all deal with variation at the infraspecific level, the three will be discussed together. Many species of plants (flowering plants, in particular) reveal considerable variation within and among populations of a single species. Classification below the species level is, therefore, frequently a challenge. Furthermore, there is still a lack of uniformity in the application of the categories of subspecies, variety, and form (Hamilton and Reichard 1992).

The term *variety* (*varietas* in Latin) was the first category to be used below the species level for plants. Linnaeus, in his *Species Plantarum*

(1753), commonly distinguished varieties, and this work was thus the beginning of even more frequent use of the term in botanical systematics. In his *Philosophia Botanica* of 1751, Linnaeus stated clearly what he believed a variety should represent: a "plant changed by an accidental cause due to the climate, soil, heat, winds, etc. It is consequently reduced to its original form by a change of soil. . . . Further, the kinds of varieties are size, abundance, crispation, colour, taste, smell. . . . Species and genera are regarded as always the work of Nature, but varieties are more usually owing to culture" (quoted in Ramsbottom 1938:199). The variety to Linnaeus, therefore, was primarily an environmentally induced variation. In modern terms this would mean a variation that was not genetically controlled in a strong or rigid way. Although Linnaeus was forceful in his definition of variety, the varieties he actually described in the *Species Plantarum* do not always coincide with his listed criteria (Clausen 1941).

The term *subspecies* was introduced into botanical systematics later than variety. The first usage of this category apparently owes to Ehrhart in a series of papers beginning in 1788 (for a summary, see Chater and Brummitt 1966a). Ehrhart regarded subspecies as "constant varieties." That is, subspecies differed from species in being separated by more minor features, but those distinguishing features tended to breed truer from generation to generation in a subspecies than in a variety. Persoon (1805) regarded the subspecies as referring to "major" morphological variations within a species whereas the variety was used only for "minor" morphological variations (Chater and Brummitt 1966b). The exact meanings of the terms "major" and "minor" are not clear, but Persoon apparently was referring to large and small differences in morphological features.

In Switzerland in the mid-1800s, Augustin Pyramus de Candolle, one of the leading botanists of the century, regarded the subspecies as the most important variation of the species. Like Ehrhart before him, he believed that characteristics of subspecies were transferred by seed from parent to offspring, but that features of varieties were less precisely controlled. In most of his publications, however, including the monumental *Prodromus* (1824–1873), he (and his son Alphonse) used the category variety more often than subspecies.

The leading taxonomist in North America in the mid-nineteenth century was Asa Gray. Gray offered this viewpoint on varieties: "Any considerable change in the ordinary state or appearance of a species is termed a *variety*. These arise for the most part from two causes, viz: the influence of external circumstances, and the crossing of races" (1836:289). No mention is made of geographical distinctness, but some of Gray's varieties were so delimited. Gray rarely used the category subspecies, but he apparently regarded it as a "marked" variety.

This early disparity in approaches to the use of the terms subspecies and variety was followed by still further differences of opinion. So great was the disagreement early in the twentieth century that some taxonomists, for example, H. M. Hall, concluded that the term variety "has such a multiplicity of uses and so often applies only to races, ecological responses, horticultural forms, or even to abnormalities that, in the opinion of the writer, its use in serious taxonomic work were better discontinued (1929:1461)." To Hall and those of similar mind (many of whom, like Hall, worked in the western United States), the category of choice was the subspecies. But many botanists, particularly those in the eastern states, took an alternative view. A representative of these eastern botanists, M. L. Fernald of Harvard, believed that the variety should be the term of choice for recognizing geographical variations of ordinary species. Subspecies, on the other hand, should be regarded as subdivisions of an aggregate species, or what we sometimes call today a *species complex*.

As a result of these historical differences within the United States on the usage of subspecies and variety, two basic schools of thought have prevailed. One is the California school, represented initially by H. M. Hall and later by Clausen, Keck, and Hiesey (1940). These taxonomists all were attempting to understand the structure of plant species by means of reciprocal transplant studies and interpopulational hybridizations. Zoologists had already worked out distinctions in terms by use of similar methods, and subspecies was the term that had stabilized in animal taxonomy. Perhaps hoping to emphasize the unity of nature, the California workers adopted the term subspecies as the primary subdivision of the species. Varieties were rarely used. The second viewpoint, promoted by taxonomists in the eastern United States, emphasizes use of the term variety for primary subdivisions of species. Subspecies are used less often, if at all.

These two opposing viewpoints for infraspecific categories gave rise to two different operations in classification at this level in the hierarchy. If morphological diversity within a species is great enough so that additional subtaxa need to be recognized, the California workers will first delimit subspecies and then recognize varieties within one or all of the subspecies if another level of distinction is needed. In approaching this same problem, the eastern workers will first delimit varieties, and then, if some of the varieties seem more similar morphologically (or in other characteristics) to each other than to the rest of the varieties, the former will be grouped into a subspecies (Kapadia 1963). In the first approach, therefore, one can have subspecies without varieties at the infraspecific level, whereas in the second approach, one can have varieties without subspecies.

There is, nevertheless, a third perspective. Some taxonomists believe that either category may be used for primary divisions of a species, the choice depending upon the degree of difference of features among the recognized taxa. In both cases, morphological differences must be associated with at least partial geographic isolation. If the difference is great, but not large enough to warrant specific segregation, then the taxa should be treated as subspecies. If the difference is small, but still represents a consistent pattern of variation, then varieties should be recognized. And if a large amount of variation exists, both categories may be used. Few workers, however, actually employ this third perspective (Hamilton and Reichard 1992).

In addition to criteria based on morphological and geographical distinctions, both varieties and subspecies are believed to be distinguishable on the basis of interbreeding. That is, closely related varieties should interbreed more often in nature or more freely in the laboratory than should subspecies, and both groups should interbreed much more freely than do species.

But what about variation that is not correlated geographically? What kind, if any, of infraspecific designations are then appropriate? It is not uncommon within natural populations to find plants with unusual morphological features growing near individuals with more "typical" morphology. These variants usually represent small genetic changes owing to mutation and/or recombination, and they are not much different in total genetic composition from the more common types. The term *form* (*forma* in Latin) is used for these unusual morphological variants, if any formal designation is believed needed. Most plant taxonomists do not use this category in a formal scheme of classification because trying to keep track of such minor morphological variations is not a useful function of biological classification. If form were attended to, goes the argument, one might as well attempt to place almost every individual organism in its own taxon because most sexual organisms are different genetically. Nevertheless, a minority of workers believe that recognizing forms serves the function of highlighting unusual morphological variations that may stimulate economic or evolutionary interest (e.g., Valentine 1975). The rebuttal, of course, is that one can indeed call attention to these variants without formal description and naming.

Genus

The next commonly used category above the species is the genus. This level in the hierarchy has a long pedigree; it was used by Aristotle to refer to a cluster of logical classes of things (usually inanimate objects) that were individually called "species." This initial use of the term *genus* for one of the more inclusive products of logical division marked the beginning of its potential application to organismic classification. Use of the concept of species for plants in general preceded the corresponding use of the concept of genus. Species in most cases were equivalent to the different "kinds" of plants distinguished by aboriginal people, but the genus was more of an abstraction. Use of a generic concept for plant classification therefore came about much later, and widespread acceptance came about later still.

Although many botanists in the Renaissance, including some of the herbalists, were grouping plant species into clusters of one sort or another, it wasn't until 1623 that Gaspard Bauhin first used the concept of genus in his *Pinax Theatri Botanici.* Bauhin, although using the concept consistently, did not provide descriptions of the genera; these were only defined by a summation of the characteristics of all the included species. Bauhin, thus, was not the founder of the modern concept of the genus. Joseph Pitton de Tournefort also used the concept of genus consistently in his *Institutiones Rei Herbariae* (1700), but in addition, he provided descriptions of each of the treated genera. For this accomplishment he is now referred to as "Father of the Generic Concept." He used monomial or polynomial generic names, which represented certain characters shared by all encompassed species, and these characters were distinctly mentioned. Linnaeus, of course, also used the concept of the genus consistently and with descriptions (e.g., in his *Genera Plantarum,* 1737). From that point on, the category of genus was well established in plant classification.

Because the genus is a higher-level abstraction than the species, the criteria by which genera are distinguished are not nearly as concrete or universal. The species relates very directly to the reproductive groupings of kinds of life on this planet. Not so the genus. A genus is a collection of species; it has a lesser biological objectivity in comparison to the species. As a result, the kinds of criteria used for delimiting genera vary depending upon the particular case. Early in the development of the concept, only morphology was available. Later came anatomy, followed in this century by cytology, cytogenetics, chemistry, etc.

The degree of morphological or other character differences that one might expect to find among genera of a particular family is thus relatively unpredictable. Genera in some families (e.g., Magnoliaceae) show striking differences in features, whereas genera in other families (e.g., Compositae, Gramineae) are separated by less obvious sets of characters. A possible reason for this range of difference, besides historical orientation within the family, is that more extinctions of species and genera have occurred in the older families, like Magnoliaceae, resulting in more marked gaps in character sets among genera.

One type of data that has been emphasized as a particularly important criterion in the distinction of genera is cytogenetics (Rollins 1953). According to the biological species concept, species are reproductively isolated from each other; both plant and animal species in general do fit this criterion. Genera, therefore, would be expected to be even more reproductively isolated from each other than are species. In my opinion, if genera cross well under either natural or artificial conditions, one should look very carefully at the species involved to be certain one or both of them have been properly classified. Of course, it is possible for some "good" genera to hybridize, and a substantial list of such hybrids has been prepared by Knobloch (1972), but this is not a common occurrence in plants.

Family and Higher Categories

Families consist of groups of genera. Some strikingly similar groups of plants that have long been recognized for their commonalities are, in fact, regarded today as families. Examples include the aromatic and square-stemmed mints (Labiatae) and the small flowers in umbel inflorescences of carrots, parsley, Queen Anne's lace, and the like (Umbelliferae). Theophrastus (370–285 B.C.) recognized these two (and several other) families many centuries ago. Formal taxonomic recognition of the term *family,* however, did not occur until after genera were recognized.

Although Tournefort had informal groups of genera in his works, he made no attempt to formally characterize these units and to provide them with descriptions. Nor did Linnaeus use a family concept consistently. Some of Linnaeus's "classes" were in fact more or less equivalent to some of our modern families, but his artificial system did not allow for a family concept to be used consistently. The use of family in the modern sense did not develop fully until the publication in 1789 of the *Genera Plantarum* by the French botanist Antoine Laurent de Jussieu. Jussieu's units (he called them "orders") correspond closely with many of the families of today, and he deliberately grouped the genera and indicated the characteristics that were used in holding them together. The term "family" was not used for these units, however, until the mid-nineteenth century.

Because the concept of the family is an even higher-level abstraction than that of the genus, still fewer criteria are available for its delimitation. As with the genus, historical precedent is an important consideration in delimiting families. And also as with genera, one would not expect species of different families to interbreed either naturally or in the laboratory. The kinds of data used to delimit families are of any and all types.

An *order* represents a group of families, a *class* a group of orders, and a *division* a group of classes. As with the development of the family concept, one would not expect the modern ideas of these higher categories to have appeared until after the family unit was employed. And in fact, it wasn't until the present century that these terms were used with any consistency. No one taxonomist is recognized as having been the first to use these three higher categories. Many people began using them, and in slightly different contexts.

Because these highest-level categories are extreme abstractions, they are the most difficult to define. Case-by-case the distinctions are usually clear, but there are no universal characters or criteria. Even today variation exists in the application of these concepts, especially in the delimitation of orders. Again, comparative data of various types are used to delimit these groups, and in some instances, strong differences of opinion prevail among plant systematists as to which families should be included in what orders. (Compare, for example, differences at the level of order in the systems of Thorne 1976, 1992 and that of Cronquist 1981, 1988). The two classes of angiosperms, monocots and dicots, are clearly differentiated by a whole series of features, but this is not always true for classes in other plant divisions.

3. Collection and Use of Data

Classification of organisms depends on comparative data for analysis. Data come in many forms, and the plant taxonomist must be able and ready to secure and handle many types of information. In addition to using data available in the literature, a taxonomist usually finds it necessary to generate original data. To fully utilize data for taxonomic purposes involves being aware not only of the taxonomic implications of the problem at hand, but also of the suitability of sampling, measurement, and display of gathered information. This chapter thus begins with a brief discussion of the types of data commonly used in plant taxonomy, followed by remarks on how to obtain and evaluate such data. For a more detailed presentation of data types and related issues (including an introduction to the relevant literature), see Stuessy (1990).

Types of Data Commonly Used

Any information about a plant or a group of plants is potentially useful for determining and understanding systematic relationships. The information already available may be detailed or sketchy, depending on the plant or taxon, and it may pertain to only one part or from many parts of the plant. From whatever source, all the different types of data can be regarded as being of two basic types: (1) information about the organism itself, such as morphology, anatomy, embryology, palynology, genetics, cytology, cytogenetics, chemistry, and reproductive biology; and (2) information about the organism-environment interactions—the ecology—which includes geographic distribution and pollination mechanisms, etc.

Both types of information are useful in taxonomic studies, but traditional work has emphasized data from the organisms themselves. Organismic rather than ecological data will, therefore, be the emphasis of this chapter. There are many kinds of organismic data, and these can be viewed as belonging to two general groups: (1) those that reflect the structural composition or architecture of the plant, such as anatomy and morphology; and (2) those that pertain to the dynamic interactions among the structures, i.e., the processes of development and physiology. Both types of data can be used in solving systematic problems. But in practice, most of the data tends to be of the structural type, owing mainly to ease of acquisition. Following is an introduction to the main kinds of structural data, along with brief discussions of reproductive biology and the general category of ecological data.

Morphology

Morphology, or the external form of an organism, has been and still is the type of data used most in plant classification. One obvious reason for this is the ease of observing and recording external features. Because of this convenience, as well as the absence of modern sophisticated equipment, the early plant taxonomists—such as Linnaeus, Jussieu, and Candolle—relied almost exclusively on morphology to classify the plants they received from collectors working in many parts of the world. The system of classification for flowering plants (angiosperms) developed by these and subsequent workers has thus been based primarily on morphological data. Both reproductive and vegetative characters have been used in angiosperm classification, but the most useful information tends to be reproductive.

The study of small morphological features, or micromorphology, has also become important in the past several decades, owing to scanning electron microscopy (e.g., Claugher 1990). Micromorphology bridges the organizational and structural gap between the external form (morphology) and internal structure (anatomy). For nearly a century taxonomists have used minute features of floral structures visible under the compound light microscope to aid in classification. In recent years, however, the scanning electron microscope (SEM) has allowed the viewing of surface features of fruits, seeds, and leaves with astonishing clarity (see several examples of such in the second and fifth case study exercises). The three-dimensional images often show many useful characters beyond the resolving power of light microscopy. The contribution made to classification by this new technology has, to date, centered on epidermal features, seed coats, and trichomes (Stuessy 1979).

Anatomy

Although anatomical features of plants have been noted for three hundred years (e.g., Grew 1682), only since the mid-nineteenth century have they been used to any significant extent in plant classification. Examples of anatomical characters include: arrangements of vascular bundles in leaves and stems, shapes and sizes of crystals, types of stomatal cells, and evolutionary modifications of floral vascularizations. Transmission electron microscopy (TEM) has proven useful for studying microanatomical features, such as variations in sieve tube-element plastids (e.g., Behnke 1986). Anatomical features do not, however, always vary a great deal from group to group (i.e., they tend to be more "conservative" characters), which can sometimes limit their utility in solving particular taxonomic problems.

Embryology

Embryological data are of more recent use in plant classification than those of morphology and anatomy. Only in the twentieth century, and in particular since the 1920s, have significant contributions been made in understanding affinities of plants based on embryological data (e.g., Schnarf 1921).

In botany, embryology usually is defined to include events and structures in the ovule prior to and after fertilization, as well as development of the male gametophyte and fertilization itself. Features used as taxonomic characters are types of ovules (orthotropous, anatropous, etc.), types of embryo sacs, and size of nucellus. Embryological data have not yet contributed nearly as much to plant classification as have anatomical features, but much more useful data are expected in the future, particularly within an evolutionary context. We know very little about the adaptive significance of embryological features.

Palynology

Palynology is the study of pollen grains and spores. Although various features of pollen have been known for more than three centuries, only since the 1830s has any significant attempt been made to describe variations in size and shape and to relate these differences to plant classification (e.g., Lindley

1830–1840). A wealth of different characters exist in pollen and spores, including number and shape of pores and wall ornamentation. Due to their small size, observations must be made with at least a compound light microscope, and SEM and TEM are routinely used.

Usually, special techniques are needed to fix the pollen or spores, section them, and prepare permanent slide mounts. The SEM tends to provide the clearest and most dramatic views of pollen grains. Because SEM technology only reveals the external appearance or sculpturing, TEM is used to reveal the internal architecture of the grains. Freeze-fracture of grains and SEM viewing is another way of obtaining external and internal data. The range of variation in pollen features within taxonomic groups may be great or small, with some taxa virtually uniform and others showing extreme variations.

Genetics

Genetics, the study of heredity, has been applied to systematic problems primarily in this century. Plant breeding, of course, has been carried on for many centuries, but use of genetic techniques in plant classification is relatively recent.

The basic methodology of the geneticist involves making artificial crosses in the greenhouse or field plot. The crosses yield hybrid generations, and the hybrid progeny may be self-crossed for advanced hybrid generations or backcrossed with either or both parents (e.g., Eggers Ware 1983). In some cases and with many additional crosses, the actual genetic basis of the characters used to separate two or more taxa can be determined. Isozyme electrophoresis is another way to assess genetic variation at the intra- and interpopulational levels (e.g., Gottlieb 1981). As these studies bear directly on isolating mechanisms and reproductive behavior of taxa, they relate to the biological species concept and therefore are most helpful at this and lower levels of the hierarchy (e.g., Crawford and Wilson 1979). The seventh case study exercise deals almost entirely with isozyme data.

Cytology

The use of cytology in plant classification is also a recent development. Not until the early part of this century were chromosomes known to be the basic units of heredity. And until the 1920s, few workers used any feature of the cell as an aid to classification. However, soon after this time, taxonomic interest turned toward the cellular level (e.g., Babcock et al. 1937). Today cytological data are routinely gathered for use in taxonomic studies.

Although any aspect of cells can be used in classification, most attention has centered on chromosomes, and on three particular aspects: number, size, and shape. Chromosomes can be viewed clearly only when the cells are actively dividing in either mitosis or meiosis. As a result, only two regions of the plant are usually sampled for analysis: (1) the shoot and root meristems (mitotic divisions); and (2) the young anther tissue (meiotic divisions in floral buds). Mother cells of eggs theoretically could be sampled also, but the few available cells make the attainment of a good preparation extremely difficult. Special killing and fixing solutions are needed to prepare properly the bud or meristem material, and compound microscopes are used for observations.

Taxonomic interpretation of cytological data requires knowledge of chromosomal evolution, which is beyond the scope of this book. However, as the case studies will reveal, taxa with the same chromosome number or with the same base (x) number of chromosomes (i.e., the lowest common denominator of higher multiples) are often closely related in the same evolutionary line. If, for example, three species suspected to share a close common ancestor are known chromosomally as $n = 10, 15$, and 20, it would be reasonable to conclude that they are all polyploid descendants of an ancestor that had a haploid number of five chromosomes, or a base of $x = 5$. The unifying factor therefore is not the individual chromosomal level of each taxon but the inferred common base number.

Cytogenetics

Cytogenetics combines the techniques of cytology with those of genetics. Taxa are artificially hybridized and chromosome pairing during meiosis is then observed within the pollen mother cells of the resultant progeny (if any can be produced). This type of data, applied initially to classification in the 1930s (e.g., Goodspeed 1934), can be very powerful in revealing relationships between selected species, subspecies, or varieties.

The assumption behind these studies is that the more closely related two taxa are, the higher the degree of pairing of chromosomes will be during meiosis. The greatest degree of pairing would result in all bivalents being formed, and the least degree with univalents (all unpaired chromosomes). Varying degrees of bivalents, univalents, and multivalents are usually encountered (see the ninth case study exercise). Cytogenetic data relate most directly to specific and infraspecific taxa and for this reason are given considerable stress in the case studies in this book.

Chemistry

The use of chemistry in plant classification is recent, having appeared in strength in the early 1960s (e.g., Alston and Turner 1963). The usefulness of chemotaxonomy was suspected in the early part of this century, but no sustained interest developed until the late 1950s. By then, technical advances had made it easier to extract, separate, and identify organic compounds.

Many types of compounds from virtually any part of a plant can be used for taxonomic comparisons. But micromolecular compounds such as phenolics (especially flavonoids; see the sixth case study exercise), terpenoids, and alkaloids have been used most extensively to date. More recently macromolecules have shown particular utility at the generic level and above (Crawford 1990).

Chloroplast DNA (cpDNA) as well as nuclear DNA are now being used extensively to assess relationships (the eighth case study exercise deals with cpDNA). Both restriction-site fragment data as well as sequence data are now important sources for taxonomic comparisons, especially in reconstructions of phylogeny. Some chemical components, such as flavonoids, are relatively easy to handle, whereas alkaloids, terpenoids, and especially DNA require more sophisticated techniques. Taxonomists and biochemists, therefore, often join together in teams to perform integrated chemotaxonomic studies. In spite of these technical difficulties in obtaining comparative chemical data, this type of information already has proven helpful in numerous studies. Chemotaxonomy will surely become even more useful in the future, especially at the macromolecular level (for a good recent overview, see Doyle 1993).

Reproductive Biology

For proper classification it is important to know not only how plants vary in vegetative and floral features but also how their reproductive systems serve to maintain these patterns. Aspects such as the timing of flowering (phenology), floral ultraviolet patterns, behavior of pollinators, nectar quantity and quality, and breeding and dispersal systems all help put into perspective the morphological, anatomical, cytogenetic, and other data. For example, if inbreeding is documented in some populations within a species, it may help explain why those populations have more uniform morphological features than have the outbreeding populations of the same species. If an entire species is inbreeding, or even commonly asexual in its reproduction, once again that taxon may reveal less overall variation in features than in a related taxon that is outcrossing.

Reproductive data tend to be efficacious in taxonomy because of the relationships of reproductive isolation and interbreeding to the biological species concept in sexual organisms. Further, because the biological species concept can be used to determine specific and infer infraspecific boundaries, and because many difficult taxonomic problems occur at this level, reproductive data are often at the core of taxonomic decisions. (Many of the case study exercises demonstrate the importance of reproductive data.)

Ecology

The ecology of plants entails two basic types of organism-environment interactions: (1) those between plants and the abiotic part of their environment (soil, temperature, moisture, etc.); and (2) those between plants and the biotic part of their environment (pollinators, predators, parasites, competitors, etc.). Both types of interactions can yield useful comparative data for systematic purposes, but ecological information, in general, has been used less frequently than the kinds of structural and physiological data already discussed.

Even so, abiotic ecological data have been used in plant classification for several hundred years. In his *Species Plantarum* of 1753, Linnaeus included observations on habitat. Today more detailed terminology exists for the description of phytogeographic areas and specific habitat types, but the same basic kinds of differences are still recognized. One recent innovation, however, is an attempt to define physiological bases for observed habitat differences among taxa (e.g., Robichaux and Canfield 1985).

The second type of ecological data, that dealing with biotic interactions, offers rich sources of information for assessing relationships. In flowering plants, for example, pollinators are sometimes species-specific, which can serve as a very useful additional character by which to distinguish species. Consider, too, the differential specificity of parasites and predators on particular hosts (e.g., Saville 1975). Many taxonomic studies have utilized such ecological features (see, for example, the second case study example).

Another type of ecological data that is frequently used in plant taxonomic studies is distribution. Distribution is simply a summary of where a plant grows geographically, usually with reference to extant political borders. In a sense, the distributional data of a species indirectly reflects both the abiotic and biotic interactions with the environment. Such data, therefore, provide very useful summaries in taxonomic studies. Distributional information is especially important at the specific and infraspecific levels because it can help reflect levels of reproductive isolation among populations.

Relative Value of Different Types of Data

In concluding this discussion of the kinds of data used for taxonomic analysis, it is pertinent to emphasize that no single type of data is supreme. For some taxonomic distinctions one kind will be determinative; for another, a different kind. For most, decisions rest on a combination. If, however, taxonomists had to choose *only one* type of data for classification, morphology would probably be the best choice.

Morphological attributes taken collectively probably give the best reflection of the overall genotype of the organisms, as well as ecological adaptations, and they are, therefore, broadly useful for purposes of classification. Fortunately, however, one rarely need make such a restrictive decision. All types of data tell something about plant relationships, and therefore as many types of data as possible should be employed. Of course, no taxonomist can be an expert in all forms of data analysis and each will have particular interests, but whether by collaboration or by broadening one's perspective, an attempt should be made to bring as many different types of data to bear on a problem as practicable. Only by this combined approach can the most useful and predictive classification result.

Gathering Comparative Data

The collection of comparative data from plants for use in taxonomic studies ought to be done in a fashion that maximizes possibilities for overall use, and with an eye to the future. It is depressing to reflect on the mass of published data dealing with cytology, chemistry, anatomy, and so forth, but which are of little use for gaining taxonomic insights because of the small sample size or lack of adequate documentation. In taxonomic studies, therefore, it is imperative to pay close attention to several considerations when gathering comparative data. Table 3.1 summarizes the points mentioned in this section. It can serve as a guide for working through the examples and exercises later in this book.

The first consideration in gathering comparative data is how one goes about sampling. This aspect of data gathering is usually done in the field, and for this reason, the investigator must have a clear idea of what he or she

TABLE 3.1.
Outline of Considerations When Gathering Comparative Data

I. *Sampling* (Field)
  A. Nature of Sample
    1. Parts of plant to sample
    2. Additional features of the plant worth nothing that will not be sampled
  B. Size of Sample
    1. Number of parts per plant to sample
    2. Number of plants to sample
    3. Number of populations to sample
  C. Techniques in Sampling: methods for collecting desired plant parts

II. *Measurement* (Laboratory)
  A. Nature of Measurement: parts of plant to measure
  B. Size of Sample to Measure
    1. Number of parts per plant to measure
    2. Number of plants to measure
    3. Number of populations to measure

III. *Data Tabulation and Graphic Display* (Office)
  A. Tabulation
    1. Raw data
    2. Numerical or statistical conversion of data
  B. Graphic Display

SOURCE: after Stuessy 1990:383.
NOTE: Four factors will realistically control the nature and extent of any data-gathering attempt: (1) purpose of study, (2) time, (3) space, (4) money.

is attempting to accomplish in the project before the field work begins. Three basic questions must be answered before sampling can be carried out effectively: (1) what parts of the plant should be sampled? (nature of sample); (2) what size should the sample be?; and (3) what techniques should be used to collect the desired plant parts?

The ability to answer satisfactorily these three questions relates directly to the ability of the worker to know where to go geographically, how much time to spend, what will be collected once there, and what kind of equipment will be needed for making the desired types of collections. These decisions, in turn, demand that the investigator have a clear idea of the type of taxonomic problem being examined, and what kinds of data will be needed to solve the problem. Some types of studies, such as those oriented primarily toward classification in a single genus, will probably require fewer populational samples and fewer individuals per population than if intensive studies are to be conducted on a close-knit complex of taxa in which hybridization is suspected.

The second consideration in gathering comparative data is measurement.

Although this aspect usually is completed in the laboratory, which gives the investigator more freedom from time restrictions, the questions regarding measurement nevertheless relate directly to those of sampling. If the field sampling is inadequate for any reason, the measurements also will be incomplete. Before a taxonomist commences field work, he or she must not only answer questions relating to sampling but also deal with these same questions in relation to measurement: (1) what parts of the plant to measure? (nature of measurement); (2) how many parts should be measured and from how many individuals in how many populations? (size of sample to be measured); and (3) what techniques will be used to obtain the desired data?

Ideally, the exact same type of equipment should be used in all cases to measure the exact same structures from precisely the same parts of the plants. For example, data on pubescence (density of hairs) should be gathered using the same rule under the same magnification and from the same organs of each plant. For measurements to be gathered in an acceptable fashion, the worker must be thoroughly familiar with the type of data being used. It must be ascertained that the techniques for obtaining the data will not cause alterations of the generated information. For example, some chemical data, such as essential oils (monoterpenoids) or isozymes, require more delicate handling. If the compounds are not treated with care (sealed vials, low temperatures, etc.), structural rearrangements can take place that might lead to a distorted view of the relationships.

The third consideration in gathering comparative data is tabulation and graphic display of the collected information. After all the data have been collected and measured, some means must be devised for reducing the data so that relationships among taxa become evident. Sometimes all that is needed is a simple listing of the raw data, but more frequently some conversion is helpful. Conversions may yield means, ranges, or standard deviations, or even more complex statistics (see Sokal and Rohlf 1981 for a good introduction to more complex methods). For the majority of taxonomists, however, a numerical treatment of data often is insufficient to illustrate the relationships clearly. A more graphic display may be necessary.

In my experience, biologists tend to divide into two groups based on their particular kind of mental abilities: (1) those who have ability to handle abstract concepts, such as mathematical or philosophical relationships that may or may not have anything to do with the world as we sense it; and (2) those who have ability to handle pattern data, particularly of the type that relates directly to objects and shapes of our experience and that can be stored as visual images. This second type of mentality is usually the kind possessed by plant taxonomists. A student is often drawn into systematics in the first place because of a strong ability to relate size, shape, and color among objects or samples and to interpret relationships. As a result of this orientation toward pattern data, taxonomists often find it useful—indeed essential—not only to tabulate the data in numerical form but also to display the data graphically. Similarly, graphic display is important for communicating the results of one's taxonomic conclusions to other workers.

Many methods can be used for such graphic display (e.g., Tufte 1983), but some of the common ones are histograms, scatter diagrams, and polygonal graphs. Other more complex methods of data display, such as isophenes, dice grams, etc., can be found in Sokal (1965) and Crovello (1970, 1974), but they will not be discussed here. When approaching the problem of data display, the most important consideration is which method (or methods) will have the strongest visual impact.

In the final analysis, four factors realistically control the nature and extent of any data-gathering effort: (1) purpose of study; (2) time; (3) space; and (4) money. Unfortunately, too often the limiting factor for our activities, whether scientific or otherwise, is money. Time and space relate directly to money, and even the purpose of a particular study is usually constrained at least indirectly by financial considerations. However, these practical considerations must never be allowed to dictate that problems be kept so narrow that no difficulty will ever be encountered in obtaining needed data. Rather, one determines first what is the problem of interest, then what the realistic requirements are. Every effort should be made to obtain the time, space, and money to gather the data needed to solve the problem.

Scrutiny, Comparison, and Evaluation of Data

After comparative data have been collected in the course of a particular taxonomic study, then they are ready for use in making taxonomic decisions. However, these data cannot be accepted and used without caution and care. The information must first be scrutinized to be certain it has been gathered properly. Only then can it safely be used to indicate discontinuities and to suggest appropriate taxa, and finally to determine ranking of the recognized groups.

The scrutiny of data involves looking closely at the generated information to be sure it is comparative in a strict sense. Three important questions must be answerable in the affirmative: (1) is the nature and size of the sample adequate? (2) have the measurements been obtained in a consistent and acceptable way? (3) are the methods of data tabulation and display appropriate? With respect to one's own data, the information should be reasonably comparative and ready for comparison and evaluation if the worker has, in fact, been attentive to these questions while gathering the data. However, if one wishes to use data obtained by other people (such as in the case studies), careful scrutiny is needed before proceeding further.

In general, a sample is adequate if it documents well the variability in characters at the next lower level in the hierarchy (whether formally recognized or not). For example, if relationships among genera are being investigated, the character variation of all the species within each genus should be known. If not all the species can be studied, this does not mean that no comparisons among the genera can be made, but the final interpretation of relationships will be less convincing. If closely related species are being examined, the infraspecific variation (at the subspecific, varietal, or populational levels) in each should be documented well before comparisons are made. In problems at any level, as new data are collected, the amount of additional variation that is added by the new samples needs to be examined. If large deviations from the inferred (or calculated) mean continue to be obtained, more data probably should be collected for a better understanding of the distribution of character states before the final relationships are assessed.

In scrutinizing the data, it is most important to know whether all the measurements were obtained in a consistent and acceptable manner. If the measurements were not properly made (ideally using the same equipment), the classifications based upon these data will be inaccurate or misleading. After the data have been scrutinized and the worker is satisfied that they are adequate for attempting a classification, the next step is to make comparisons among the data with the ultimate aim of constructing taxonomic groups (or taxa) and ranking them. The primary basis for delimiting taxa is the recognition of discontinuities. Discovery of discontinuities, in turn, depends on the recognition of similarities and differences. It is assumed that the more closely related two individuals or taxa are, the more similar will be all the data obtained from them. Therefore, reasoning backward, when approaching a new set of individuals or taxa, those that have more and different types of data in common are assumed to be the most closely related.

Finally, methods of data tabulation and display should be suitable for the problem at hand (Tufte 1983). This aspect is crucial for communicating a view of relationships to someone else. In fact, it is no exaggeration to state that without effective data tabulation and display (especially visual display), it is unlikely that others will be convinced of the utility or merit of one's classification. Therefore, after examining relevant data, pains must be taken to choose a method of display for part (or all) of them that show most clearly the relationships that the worker believes exist.

The Three Schools

At present there is no agreement among taxonomists as to the most appropriate method for comparing data for purposes of classification. In fact, three distinct schools exist: phenetics, cladistics, and phyletics. The phenetic school advocates use of many unweighted characters to form groups. It is highly quantitative. No initial distinction is made as to whether one character (say, number of petals) is more important than another (say, color of petals). The logic underlying this approach is that the sheer quantity of characters will yield a satisfactory result. The 1973 book by Sneath and Sokal is a good exposition of this school.

The cladistic school stresses the selection and use of a smaller number of evolutionarily significant characters, with shared derived character states (= synapomorphies) being most important for group formation. Like the pheneticists, cladists believe that their approach diminishes the bias of worker intuition. It is also quantitative. But similarities end there. Cladistics differs from phenetics in that the former views all aspects of classification within an evolutionary context, whereas the latter de-emphasizes the evolutionary import of characters, states, and trees of relationship. Cladistics employs fewer characters for which evolutionary directionality of states (= polarity) can be determined with some confidence. The 1966 book by Hennig is a seminal document for this school.

The phyletic (or evolutionary) school emphasizes all aspects of phylogeny in reaching a final classification. It concentrates on evolutionarily significant characters for reconstructing branching patterns of phylogeny, but in addition it takes into account the degrees of divergence within lineages. That is to say, for purposes of classification, it is not only important to know common ancestors, but also the amount of evolutionary change that has occurred since the initial divergence. Although phyletics has been done for more than a century by intuitive (i.e., nonexplicit) means, several quantitative methods now also exist (Estabrook 1986; Stuessy 1987; Ashlock 1991; Hall 1991). The most important advantage of phyletics over cladistics is the greater amount of evolutionary information contained in the final classification. (For detailed comparisons of these three schools, see Stuessy 1990).

Whichever school a student may choose to align with and whatever methodology may be best for a particular problem, I nevertheless firmly believe that a solid grounding in the kinds of judgments employed with the intuitive phyletic approach is invaluable. Hence the subtitle of this book: "exercises in applied pattern recognition." The case study exercises in this book have been selected to give the student experience in dealing with character data and in making intuitive judgments about their efficacy in particular situations. The more individual judgment that must be brought to bear on a problem, the more the student learns about effective classification.

Understanding Organic Evolution

The ultimate cause for discontinuities among organisms in nature lies with organic evolution. In a general sense, life is a continuum from small and simple organisms to those that are large and complex. If any part of this continuum is examined closely, however, it is seen to consist of many gaps. Discontinuities prevail.

Two evolutionary processes are largely responsible for these discontinuities: speciation and extinction. Speciation is the mechanism (or mechanisms) by which new species are formed. Whenever speciation occurs, the new species is discontinuous in regard to the attributes of the progenitors, and this discontinuity usually is maintained by reproductive isolation—a populational phenomenon. Studies of patterns of variation, and reasons for these patterns, such as hybridization, breeding systems, pollinator specificities, and ecotypic differentiation, impact directly on concepts of species and infraspecific taxa. An understanding of these populational processes, therefore, can greatly enhance our ability to make predictive classifications at these levels in the hierarchy. Extinction can also create discontinuities in data sets, especially if some of the intermediate taxa are lost and no fossils are known. Without connecting data, the taxa seem more discontinuous from each other than they would have appeared before extinction took place.

Because of this relationship between the evolutionary process and the existence of discontinuities in nature, it is obvious that an understanding of organic evolution in plants (including fossil evidence and populational variability) is highly desirable for effective modern plant classification. Imparting such an understanding is beyond the scope of this book, but the interested reader is referred to Stebbins (1977) for a good, though brief, overview and to Grant (1981) for more details.

4. How to Determine Taxonomic Relationships

Having completed a brief discussion on classification in the first three chapters, it is now appropriate to turn our attention to solving specific taxonomic problems. The general way to solve any problem of classification involves examining the gathered data with reference to the levels in the hierarchy at which the problem resides. Foremost, we must pay attention to two things: (1) correlations of character states among taxa; and (2) discontinuities in character states between taxa.

At a practical level, therefore, the way to make sense of any taxonomic problem is to determine the distribution of character states over all taxa under consideration, beginning with those character states that historically have given good results and in which obvious variation exists (e.g., one would more probably look first at chromosome numbers than leaf blade length). Consideration of each character will allow one to divide up the taxa into subgroups. In general, the best interpretation of taxonomic relationship is that for which the majority of characters correlate, thereby suggesting the same groups.

In addition to morphological character states, geographic distribution may also reflect taxonomic relationships, especially at the specific and infraspecific levels. So may phenology (flowering time), type of pollinators, breeding systems, and so forth. The ability of populations to interbreed and produce hybrid offspring of various degrees of fertility is also important.

This chapter prepares the way for the case study examples and exercises by providing simple hypothetical examples and problems. Readers may thus begin to gain confidence in making taxonomic decisions. First, the hypothetical examples demonstrate how one goes about recognizing taxonomic groups by seeking correlations among character state distributions. The hypothetical problems that follow are more complex, as they involve data typically used in taxonomic studies. Concluding the chapter is a discussion of these examples and problems, aimed at teaching the rudiments of grouping and ranking with typical data. The problems all refer to angiosperms (the flowering plants), as do the case studies in chapters 5 and 6, but the approach to problem solving would be the same for most other vascular plants, and to a lesser degree for most other organisms.

Recognizing Patterns

The methodology of and difficulties involved with recognizing groups can be illustrated by a few simple examples. Figure 4.1 shows examples involving five hypothetical taxa (A–E). Example 1 in figure 4.1 is a situation in which only one datum (or character) is known for each taxon, and this character has only two observed and recorded states (+ and −). In this simple example, an obvious discontinuity exists among the data so that A and B are most similar (closely related) and C, D, and E also cluster together. Example 2 is slightly more complex, in that another set of data (another character) has been added, which also occurs in only two states. Here the data correlate absolutely and show strong and unequivocal discontinuities, so that A and D are placed in one group and B, C, and E in another. Example 3 illustrates the same type of problem with three characters. This time three distinct groups prevail, based on correlations among the three characters: A and E; B and D; and C by itself. Taxon C is slightly more similar to B and D than to A and E, but C is also more dissimilar from B and D than the latter are from A and E. Three units seem justified.

The next three examples (4–6) show the same number of taxa but with increasing numbers of characters. Example 4 is complex, with more difficulties in forming groups than one encounters in example 5, which has an additional character. Example 6 illustrates quite a difficult situation, with taxon D having very strong affinities with A, B, and E, and only slightly less

FIGURE 4.1 Six examples of how to recognize discontinuities in data with increasing numbers of characters. Plus sign indicates presence of a particular character state; minus sign indicates presence of a contrasting state.

Taxa	Character 1
A	+
B	+
C	–
D	–
E	–

Classification: group A, B
group C, D, E

1

Taxa	Characters 1	2
A	–	–
B	+	+
C	+	+
D	–	–
E	+	+

Classification: group A, D
group B, C, E

2

Taxa	Characters 1	2	3
A	+	+	–
B	–	+	–
C	–	–	+
D	–	+	–
E	+	+	–

Classification: group A, E
group B, D
group C

3

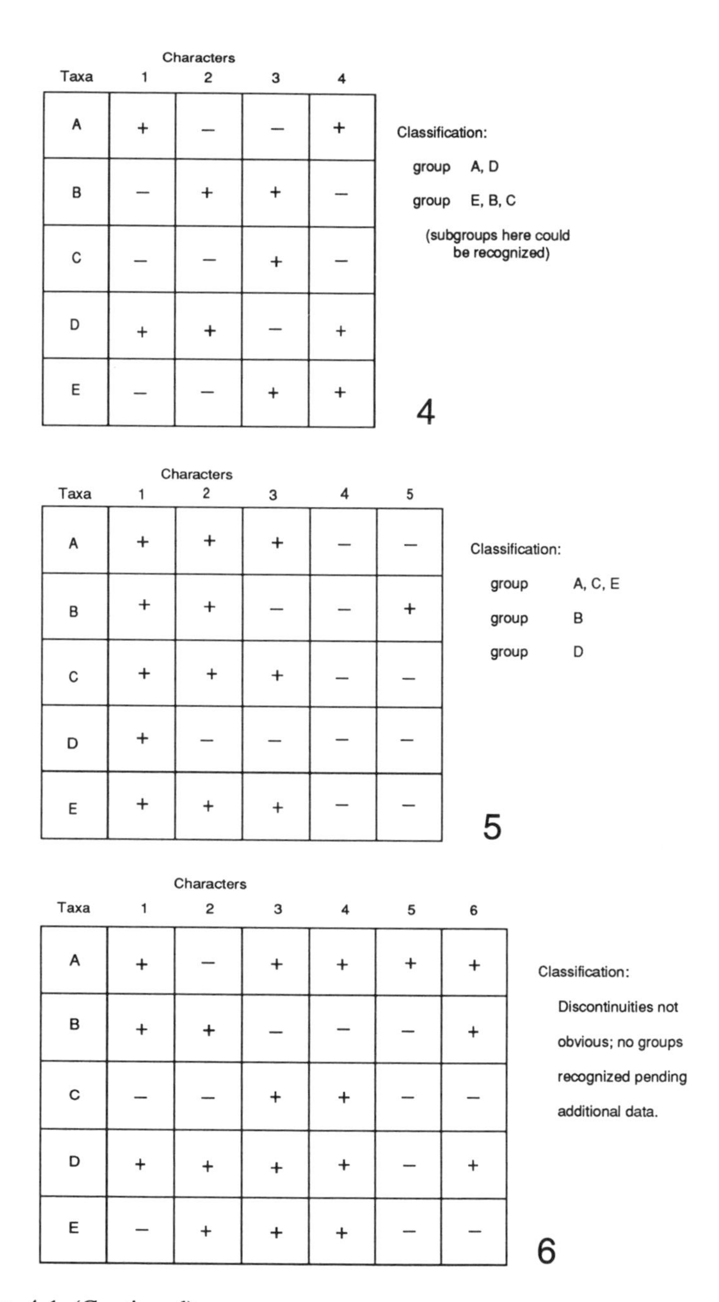

Taxa	Characters 1	2	3	4
A	+	−	−	+
B	−	+	+	−
C	−	−	+	−
D	+	+	−	+
E	−	−	+	+

Taxa	Characters 1	2	3	4	5
A	+	+	+	−	−
B	+	+	−	−	+
C	+	+	+	−	−
D	+	−	−	−	−
E	+	+	+	−	−

Taxa	Characters 1	2	3	4	5	6
A	+	−	+	+	+	+
B	+	+	−	−	−	+
C	−	−	+	+	−	−
D	+	+	+	+	−	+
E	−	+	+	+	−	−

FIGURE 4.1 (Continued)

strong relationship with C. In this last example, the data do not show marked correlations or discontinuities, and it would be difficult to provide a convincing classification. The most reasonable approach in this case is simply to treat all the taxa together as a unit until additional data might suggest a useful partitioning.

These simple examples illustrate two important points about the practical aspects of classification. First, it is usually true that the more characters and taxa that are involved, the more complex becomes the recognition of groups. Sometimes, however, the taxa are relatively obvious even though a large data matrix is used (e.g., example 5 in contrast to example 6). Second, sometimes the data in hand, irrespective of the number of taxa, simply do not sort well into groups (e.g., example 6). In this situation more data are needed before formal recognition of taxa is advisable.

Whereas the preceding six hypothetical examples provide insights for the practical methodology of data comparison, the amount of data is less and number of taxa are fewer than what one typically encounters in actual taxonomic problems. To provide an even more extensive introduction to the problems associated with plant classification, therefore, three hypothetical problems are presented and discussed next.

Delineating Groups: Three Hypothetical Problems

The first hypothetical problem is illustrated by the data in table 4.1. The arrangement of these ten taxa into units based on similarities and differences is very simple. In regard to habit: A and E are herbaceous; B, G, and H are shrubs; and C, D, F, I, and J are trees. This simple piece of data about the habit of the plants, therefore, initially divides the taxa into three groups. The next step is to see if additional data correlate with these habit differences. The position of the ovary in the flower does correlate exactly with the differences in habit: A and E are hypogynous; B, G and H are epigynous; and C, D, F, I and J are perigynous. At this point each of the three groups becomes more cohesive within itself and more distinct from the other groups. The third type of data provides further corroboration of the recognition of these three groups. A and E have chromosome numbers on a multiple of 10; B, G, and H are multiples of 6; and C, D, F, I, and J are all based on 7. All data correlate, therefore, to suggest that three good groups exist among the ten taxa. It makes no difference in this case which character is analyzed first because all correlate to give the same groups.

TABLE 4.1.
Data for the First Hypothetical Problem

Taxon	Habit	Flower Type in Relation to Ovary Position	Chromosome Number (n)
A	herb	hypogynous	10
B	shrub	epigynous	6
C	tree	perigynous	7
D	tree	perigynous	7
E	herb	hypogynous	20
F	tree	perigynous	28
G	shrub	epigynous	12
H	shrub	epigynous	18
I	tree	perigynous	7
J	tree	perigynous	14

The second hypothetical problem is illustrated by the data in table 4.2 and figure 4.2. Turning first to table 4.2, we can attempt to group the taxa into units based on morphological and cytological similarities and differences. No simple method for doing this exists, but often a good way to proceed is to compare individual features and make a list of the observed groups that result. Then one can look for the same groups of taxa reappearing in all the other features.

Four groups thus emerge in this hypothetical problem: taxa A, F, and I clump together, as do B, C, G, and H; D and J stand together, and E is left alone. All data sets do not show the same relationships, however. For example, the longevity, sepal number, and ovary position in the species each show only two character states, whereas seven different character states denote flower color. Nevertheless, leaf shape, leaf vesture, stamen number, and chromosome number all correlate to give four reasonably distinct groups. Several discontinuities are, however, worth noting. Taxon G in the group B, C, G, and H is perennial and the rest are annual. This same taxon has a petal color of orange, which is intermediate between the red color of B and C and the yellow pigments of H.

Reproductive crossing relationships among all ten taxa in this problem are shown in figure 4.2. The group consisting of taxa A, F, and I (which were already interpreted to be strongly related in morphological features) do cross and leave progeny at a success rate of 76–100 percent. Breeding relationships thus support the conclusion of relatedness for these three taxa, based on morphological features. D and J also cross, if weakly; and B, C, G, and H are all connected variously by strong to weak crossability. Taxon E crosses

TABLE 4.2.

Morphological Features and Chromosome Numbers of Taxa in the Second Hypothetical Problem

Taxon Number	Longevity	Leaf Shape	Petal Color	Sepal Number	Stamen Number	Ovary Position	Leaf Vesture	Chromosome Number (*n*)
A	perennial	ovate	magenta	5	10	inferior	strigose	9
B	annual	oblong	red	4	many	superior	sericeous	5
C	annual	oblong	red	4	many	superior	sericeous	5
D	annual	obovate	blue	4	8	superior	stipitate-glandular	16
E	perennial	lanceolate	white	5	5	superior	peltate	12
F	perennial	ovate	pink	5	10	inferior	strigose	9
G	perennial	oblong	orange	4	many	superior	sericeous	5
H	annual	oblong	yellow	4	many	superior	sericeous	5
I	perennial	ovate	pink	5	10	inferior	strigose	9
J	annual	obovate	blue	4	8	superior	stipitate-glandular	8

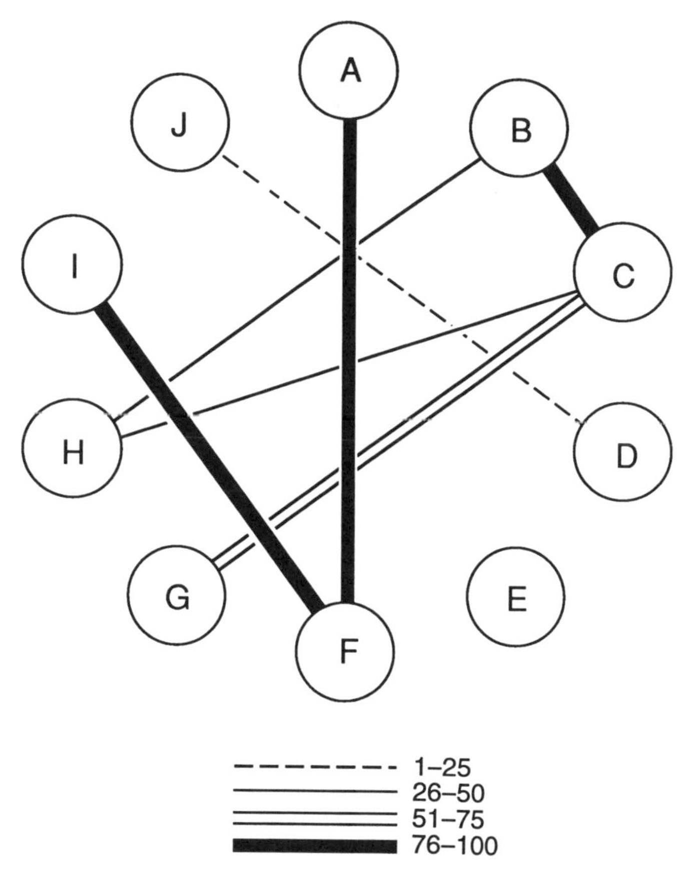

FIGURE 4.2 Diagram of reciprocal crossability relationships among all taxa in the second hypothetical problem. Lines connecting taxa indicate viability of F_1 progeny obtained (as percentage of total flowers used in the cross). An absence of connections between taxa indicates that no viable progeny were obtained in those particular crosses.

with nothing else. The crossability data, therefore, give corroborating evidence that four groups should be recognized: (1) E; (2) A, F, I; (3) B, C, G, H; and (4) D and J.

The third hypothetical problem is illustrated by the data in tables 4.3 and 4.4 and figures 4.3–4.5. To begin, table 4.3 lists the morphological and chromosomal features of the ten taxa under study. In contrast to the two previous problems, both quantitative and qualitative characters are given for morphological features. Because the quantitative values are presented as

TABLE 4.3.

Morphological Features and Chromosome Numbers of Taxa in the Third Hypothetical Problem

Taxon	Plant Height (cm)	Leaf Shape	Leaf Margin	Leaf Vesture Lower Surface	Corolla Diameter (mm)	Leaf Length (cm)	Petal Color	Chromosome Number (*n*)
A	20–30	ovate	entire	tomentose	9–12	12–14	red	7
B	35–50	narrowly ovate	entire	tomentose	11–13	13–16	red	7
C	20–40	linear	dentate	sericeous	5–9	5–7	yellow	6
D	40–60	ovate	entire	tomentose	8–12	14–20	red	7
E	10–20	elliptic	serrate	strigose	11–18	6–7	orange	9
F	10–20	elliptic	serrate	pilose	16–22	6–7	orange	9
G	50–70	obovate	doubly serrate	hirsute	26–31	8–12	orange	8
H	30–50	linear	dentate	sericeous	7–10	6–8	yellow	6
I	70–85	obovate	doubly serrate	hispid	20–26	10–15	orange	8
J	80–100	obovate	doubly serrate	hispid	19–28	10–14	orange	8

NOTE: Quantitative values are ranges obtained from measurements of at least ten individual plants found throughout the distribution of each taxon.

TABLE 4.4.
Occurrence of Phenolic Compounds from Leaves of Taxa in the Third Hypothetical Problem

	Compound Number											
Taxon	1	2	3	4	5	6	7	8	9	10	11	12
A	X	X	X	X	X	X	X					X
B	X		X	X			X					X
C								X	X	X	X	
D		X	X	X	X	X	X	X			X	X
E	X								X		X	
F	X	X							X		X	
G	X								X			
H								X	X	X	X	
I	X	X							X			
J	X	X							X			

NOTE: A compound was regarded as present (X) in a taxon of it occurred in over 50 percent of the individuals analyzed.

ranges that overlap from taxon to taxon, it is better to begin the evaluation with the qualitative characteristics.

As in the previous example, it is best to take each character separately and see if groups of taxa result, then to look for correlations of groupings that should indicate relationships. After analyzing the data in table 4.3, four groups are clearly seen: A, B, and D; C and H; E and F; and G, I, and J. These all might be equivalent taxonomic units, except that they are not equally different from each other. For example, the group of E and F is, in this way, closer to the group of G, I, and J than A, B, and D is to group C and H.

As a working hypothesis we may, nevertheless, regard these four groups as equal, but we should be cautious. More data obviously are needed. The quantitative values now can be consulted to see if they correlate with the four tentative groups already established via qualitative features. Basically, these additional data are ambiguous because of the overlapping ranges. One relationship that shows clearly, however, is the closeness of E and F. C and H also seem very close, as do I and J.

The phenolic chemical data are shown in table 4.4. Phenolic compounds are secondary plant products that can be extracted from leaf tissues and analyzed chromatographically. If we take the four groups already indicated by morphologic data and see how they correlate, we find that the group of A, B, and D reappears in the chemical data, as do the groups C with H and E with F. Taxa I and J pair again, but G can be just as well put with the E/F

TABLE 4.4
(*continued*)

Compound number												
13	14	15	16	17	18	19	20	21	22	23	24	25
X	X											
	X	X										
									X	X		
X		X										
			X	X	X	X	X	X			X	
			X	X	X	X		X			X	X
								X			X	
									X	X		
	X	X	X	X				X			X	
	X	X	X	X				X			X	

group as with the I/J group (with which it is correlated morphologically). We need more data.

Turning to figure 4.3, we see the viability of pollen in the F_1 hybrids obtained during attempts to cross all taxa. A reasonable assumption is that the more fertile the pollen are in the hybrid, the more closely related are the parental taxa. Taxa A, B, and D cross well, and the hybrids have 26–75 percent viable grains. C and H also cross well and the hybrids are very fertile (76–100 percent). Taxa E and F cross well, too; their hybrids have 51–75 percent fertility. Notably, G, I, and J all cross with each other at the 51 percent level or above—thus easing the uncertainty about the affinities of G raised by the chemical data. The data from artificial hybridization thus substantiate the four groups we initially delimited morphologically.

There is, however, one more set of data we can evaluate for this third hypothetical problem: geographic distribution. The distributions of the ten taxa overlap to a large degree (figures 4.4 and 4.5); these data are not, therefore, particularly useful for delimiting groups. Geographic data will be useful, however, in the ranking of the taxa (to be discussed in the next section).

In these three hypothetical problems, the taxa did sort satisfactorily into groups. Most of the time actual taxonomic problems also resolve satisfactorily. Sometimes, however, no well-delimited groups are discovered. If this happens, usually one or more of three factors are responsible: (1) not enough data or different types of data are at hand; (2) the taxa involved really belong

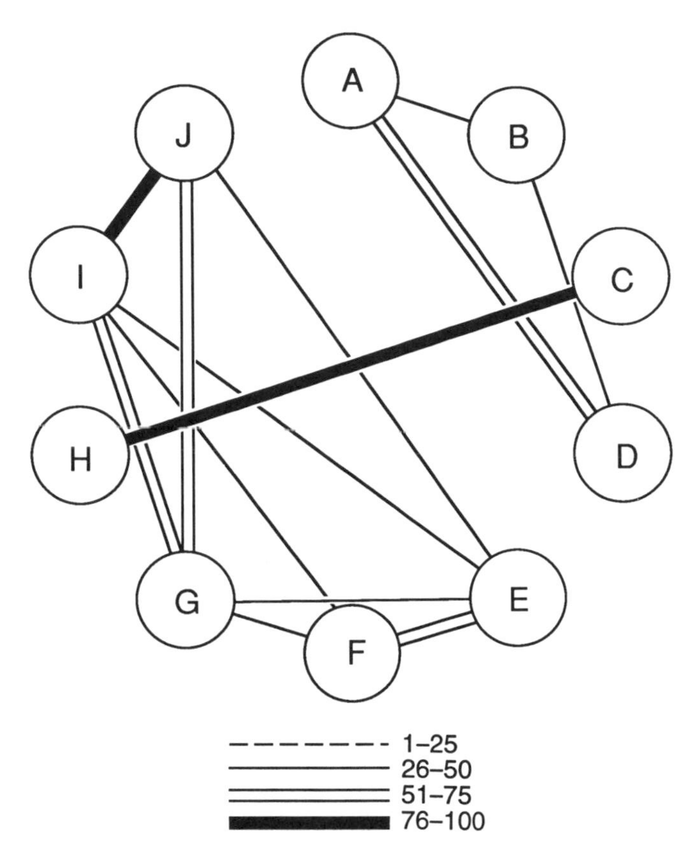

FIGURE 4.3 Pollen viabilities in F_1 hybrids (as percentage of viable grains) produced by reciprocal crosses among taxa of the third hypothetical problem. An absence of connections between taxa indicates that no hybrid progeny were obtained in the cross.

to the same group and will never partition, even with more data; and (3) the dynamics of the evolutionary process, such as hybridization or ecotypic differentiation, are obscuring the discontinuities. A good knowledge of the evolutionary process, therefore, enables a taxonomist to approach more effectively and interpret more successfully the existing patterns of relationships in these more challenging situations. While it is true that a person does not have to know anything about evolution to be a good taxonomist, it is also true that the acquisition of such additional knowledge will make an even better worker.

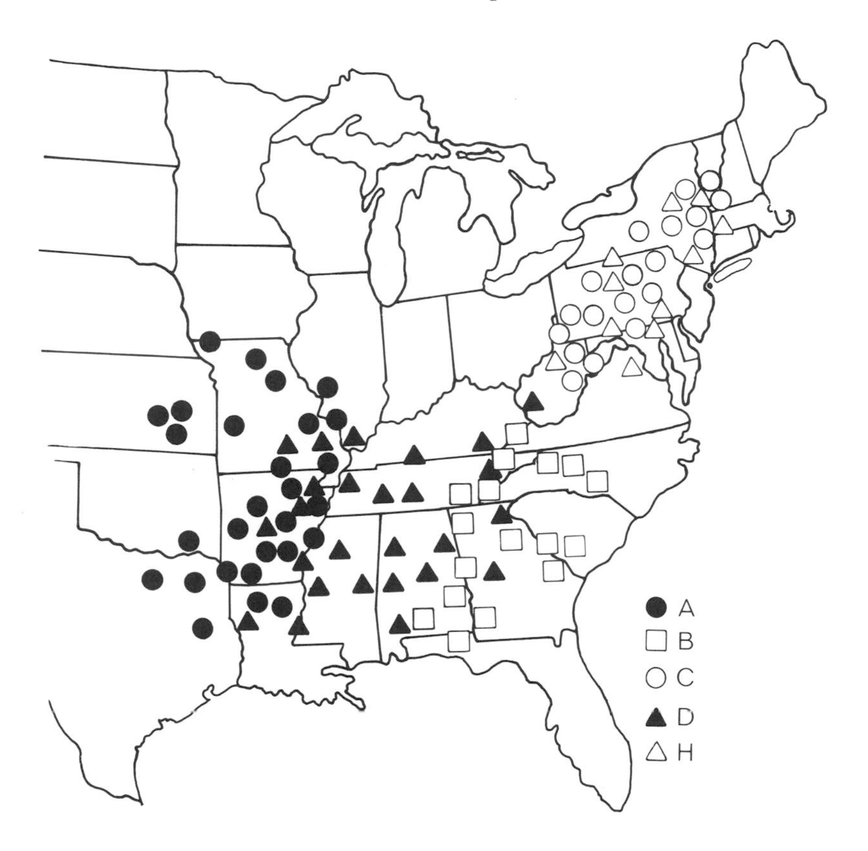

FIGURE 4.4 Known distributions of taxa (A, B, C, D, H) in the United States in the third hypothetical problem.

Ranking the Sorted Groups

Having considered in some detail the methodology of comparing different sorts of data, we need to turn to the evaluation of the data for purposes of ranking. With any system of classification (phyletic, phenetic, cladistic, or even artificial), ranking the resultant groups (or taxa) depends on an evaluation of the characters and character states. This evaluation is done by constructing a hierarchy of the taxa in a way that emphasizes the similarities and differences (or the kinds and degrees of discontinuities) among the groups.

As an example of the methodology of ranking, let us return to the third hypothetical problem. Four groups were recognized: (1) A, B, D; (2) C, H;

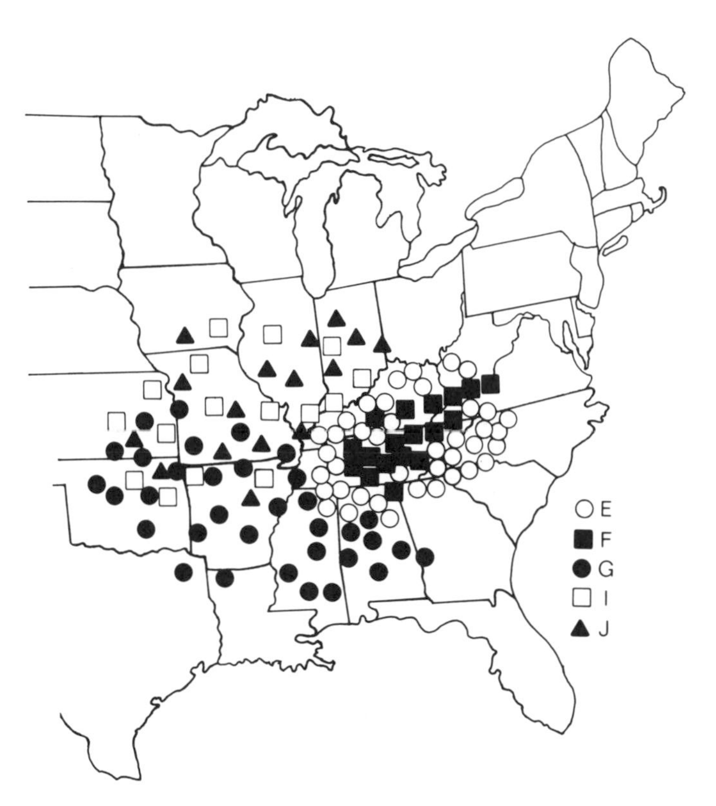

FIGURE 4.5 Known distributions of taxa (E, F, G, I, J) in the United States in the third hypothetical problem.

(3) E, F; and (4) G, I, J. Although four groups were delimited, these four were not all equally discontinuous with each other. For example, group E/F seems closer to group G/I/J than it does to groups A/B/D and C/H, especially in the pollen viability of the F_1 hybrids. If this is the case, then groups E/F and G/I/J are perhaps best regarded as a single larger group that can be internally subdivided into two subgroups. We also need to explore the relationships within each of these subgroups.

By looking at the quantitative and qualitative morphological differences more carefully, we notice that E and F are very similar in all respects, except F seems to have pilose leaf undersurfaces and larger corollas (table 4.3). In phenolic chemistry, the two differ in only a few compounds (table 4.4). Geographically (figure 4.5), both E and F are confined more or less to the southern part of the Appalachian Mountains of the southeastern United

GROUP 1	GROUP 2	GROUP 3
unit 1 Taxon A	subunit 1 Taxon C	subgroup 1
unit 2 Taxon B	subunit 2 Taxon H	unit 1 Taxon E
unit 3 Taxon D		unit 2 Taxon F
		subgroup 2
		unit 1 Taxon G
		unit 2
		subunit 1 Taxon I
		subunit 2 Taxon J

FIGURE 4.6 Ranking of taxa within the third hypothetical problem.

States, but F is found mostly at the higher elevations in this mountainous region. Based on all data, one might treat E and F as coordinate units of one subgroup, i.e., they belong together at the same level in the hierarchy.

The taxa of the other subgroup (G, I, and J) can, however, be further classified. I and J are nearly identical in morphological, cytological, and phenolic chemical data. In addition they cross with great ease. Distributionally (figure 4.5) they are intermixed throughout the central states and, as a result, seem best treated as a subunit within the larger group, the two standing together as a single unit paired with G at the next level. On the other hand, within the group A/B/D each of the taxa is more or less equidistant from the other in various features. Therefore, each seems deserving of unit (or subgroup) status. Meanwhile, C and H show about the same minor differences from each other as do I and J, and the former are thus best treated as subunits or as belonging to the same group without internal classification.

Out of all this comparison in relative differences among the components of the established groups we can derive the ranking displayed in figure 4.6. In this ranking, group represents the highest category, subgroup the next, followed by unit, and then subunit. Note that only Group 3 has subgroups and that the distinction between taxon C and H in Group 2 is so small that they must be delineated as subunits rather than units.

At this stage we have grouped and ranked the ten taxa, and a useful hierarchy of classification has been generated. However, this hierarchy has limited biological meaning. That is, the arbitrarily selected categories in our hierarchy (group, subgroup, unit, and subunit) do not relate in any way to those of the Linnaean hierarchy (genus, species, variety, etc.). Our next step, therefore, is to place the taxa in equivalent categories of the Linnaean hierarchy. Because the species is the most clearly defined unit, it is wise to begin working at this level and then continue upward to the generic and downward to the subspecific and varietal levels.

According to a strict interpretation of the biological species concept, we would expect populations of a species to be similar in all data sets as well as being reproductively isolated from populations of other species. Returning to figure 4.3, we can see that our group 1 (A, B, and D) crosses well within itself, but will not cross well with groups 2 or 3. Similarly, group 2 (C and H) crosses very well within itself, but not at all with groups 1 or 3. And finally, group 3 (E, F, G, I, and J) crosses within itself but not with groups 1 or 2. Because these three groups are morphologically and chemically similar within themselves but discontinuous with each other and because each is reproductively isolated, it is likely that each group represents a biological species. Our informal category "group," therefore, can now be changed to the category "species" in the Linnaean hierarchy. As this is the broadest level of grouping in our hierarchy, the generic level is not in question, and we can guess that the other informal categories (subgroup, unit, and subunit) probably relate in some fashion to subspecies, varieties, or forms in the Linnaean hierarchy.

Within species 1 (the old group 1), all three "units" have ranges of distribution in the southeastern United States that are reasonably distinct from one another, although A overlaps with D, and D with B. Because of this geographical separation and the degree of discontinuities among the taxa, one might wish to recognize them as three varieties of one species. As mentioned earlier, workers of the California school or Europeans might prefer to call these units subspecies instead of varieties.

SPECIES 1		SPECIES 2		SPECIES 3	
variety 1	Taxon A	form 1	Taxon C	subspecies 1	
variety 2	Taxon B	form 2	Taxon H	variety 1	Taxon E
variety 3	Taxon D			variety 2	Taxon F
				subspecies 2	
				variety 1	Taxon G
				variety 2	
				form 1	Taxon I
				form 2	Taxon J

FIGURE 4.7 Linnaean hierarchy of taxa within the third hypothetical problem.

Within species 2 (the old group 2), C and H are so similar to each other in all respects, including complete overlapping of distributions, that they should be treated not as varieties or subspecies but as "forms"—if they are recognized formally at all.

Within species 3 (the old group 3), we have the most diversity. E and F have geographic ranges distinct from G, I, and J. Furthermore, F is confined to central and eastern Tennessee and adjacent regions, whereas E surrounds the populations of F. As for crossing relationships, hybridization between taxa of subgroup 1 and 2 results in hybrids with only 26–50 percent pollen viability, substantially reduced from the intrasubgroup crosses. Because of this large degree of discontinuity, it seems prudent to recognize each of the subgroups as distinct subspecies—not varieties, and certainly not forms. Within the first subspecies, taxa E and F cross moderately well, but they have separate ranges of distribution that overlap only at the margins. These might be recognized best as varieties. Within the second subspecies, it is obvious that I and J are closer in their crossability than either is to G. Similarly, the distributions of I and J are completely overlapping. One might call them forms, if they are recognized formally at all. The only taxon left, G, has a reduced viability of pollen in hybrids with I and J and also has a relatively distinct distributional pattern. These kinds and degrees of discontinuities seem indicative of varietal recognition. Our new classification utilizing categories of the Linnaean hierarchy is shown in figure 4.7.

5. Case Study Examples

It is now time to examine how actual taxonomic decisions are made using three real examples from botanical journals. The data in these real examples are more complex than in the hypothetical cases, and hence more challenging. The types of problems represent situations that are frequently encountered in plant classification, with particular emphasis on specific and infraspecific categories. Each of the three case study examples is organized into five sections: (1) statement of the problem; (2) comments on materials and methods used in the data collection; (3) presentation of data in the form of figures and tables; (4) citation of the original paper from which the data were taken (title deliberately omitted); and (5) discussion of the way the original author(s) treated the problem. It is hoped that by carefully following and working through the examples in this section, the reader will be prepared to attack the case study exercises in chapter 6, which will, of course, omit the discussion section.

Case Study Example 1 (Compositae)

PROBLEM: Provide a classification for the taxa A–D using any or all of the following categories: genus, species, subspecies, variety, form.

MATERIALS AND METHODS: The taxa under study have been collected rather sparingly. Specimens borrowed from 12 herbaria yielded a total of only 72 collections. This material was supplemented by field observations and mass collections from 13 populations. (See figure 5.1.1 for locations of specimens collected.)

Buds collected in the field for chromosomal studies were fixed in a modified Carnoy's solution (4 parts chloroform: 3 parts absolute alcohol: 1 part glacial acetic acid) and refrigerated immediately. They were transferred after 24 hours to a solution of 70% ethyl alcohol "colored" with iron acetate and refrigerated until squashes were made some months later. An aceto-orcein stain (0.5% orcein in a solution of 1 part glacial acetic acid: 1 part lactic acid: 1 part water) was employed. Mitotic counts were made from germinating seeds following colchicine treatment.

A single count of $n = 12$ for Taxon B has been reported by Raven and Kyhos (1961. *Amer. J. Bot.* 48: 842–850). This was corroborated by both meiotic and mitotic counts for plants in population *698*. Meiotic counts of $n = 12$ were obtained for populations *700, 703,* and *705*—all of these being referable to Taxon A. Meiosis was regular in all pollen cells examined except for occasional bridges at first anaphase in population *703*. Bivalents ordinarily formed two terminalized chiasmata at late diakinesis.

After thorough morphologic investigation of all 13 populations surveyed in this study (see figures 5.1.3 and 5.1.4 for drawings of each taxon), certain characteristics of stems and leaves were found to be of greater diagnostic value than others in rendering taxonomic judgments. Figure 5.1.2 presents quantitative measurements of these characters. For these data no less than 20 individuals from each population were measured. One typical lower leaf per stem was used for blade width, length, and length/width measurements. The total upper surface of these leaves was examined with care in order to determine the minimum and maximum trichome lengths. The single longest leaf per individual stem was utilized for maximum leaf length, and all stem heights were accounted for in attempting to evaluate the variation in the size of the plants. Table 5.1.1 supplements the data presented in figure 5.1.2. In it is summarized the remaining most significant morphologic variation that was found.

DATA: figures 5.1.1 through 5.1.4; table 5.1.1.

REFERENCE: Ellison, W. L. 1971. *Brittonia* 23: 269–279.

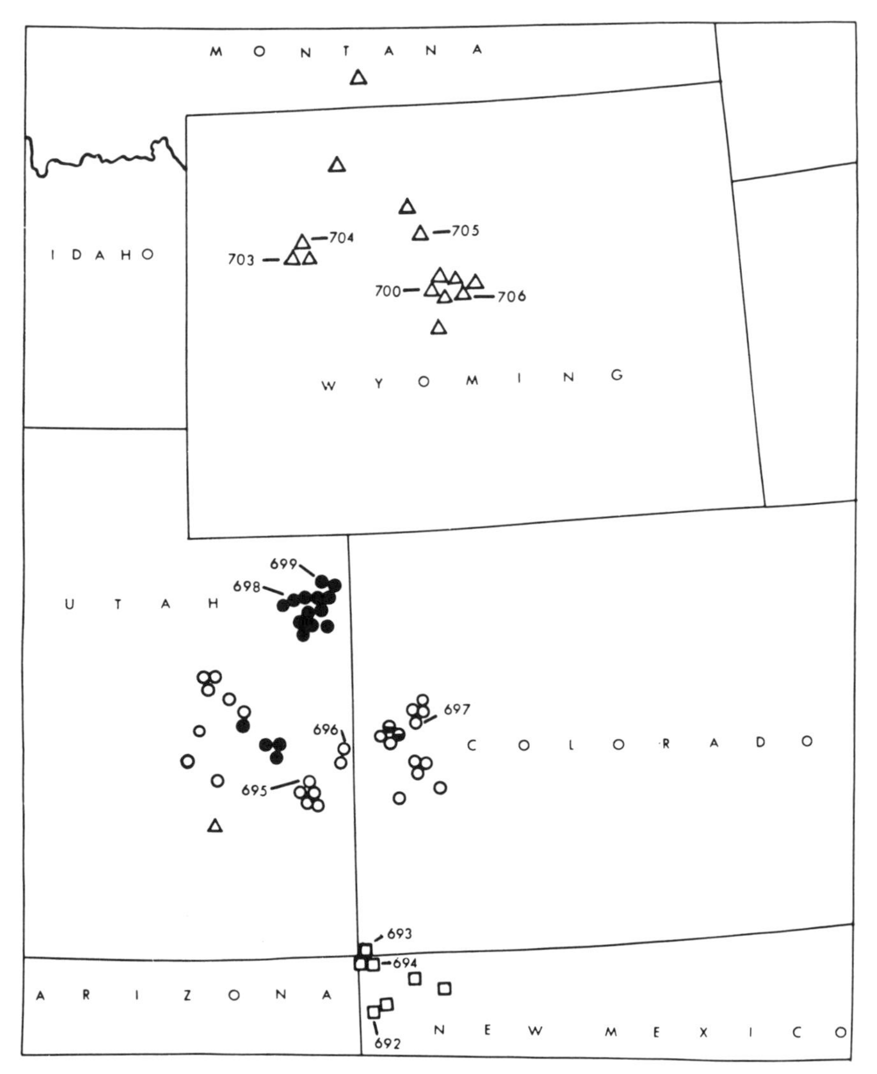

FIGURE 5.1.1. Map showing distributions of taxa A–D. Taxon A, triangles; Taxon B, blackened circles; Taxon C, circles; Taxon B–Taxon C intermediates, half-blackened circles; Taxon D, squares. Numbers indicate locations of mass collections.

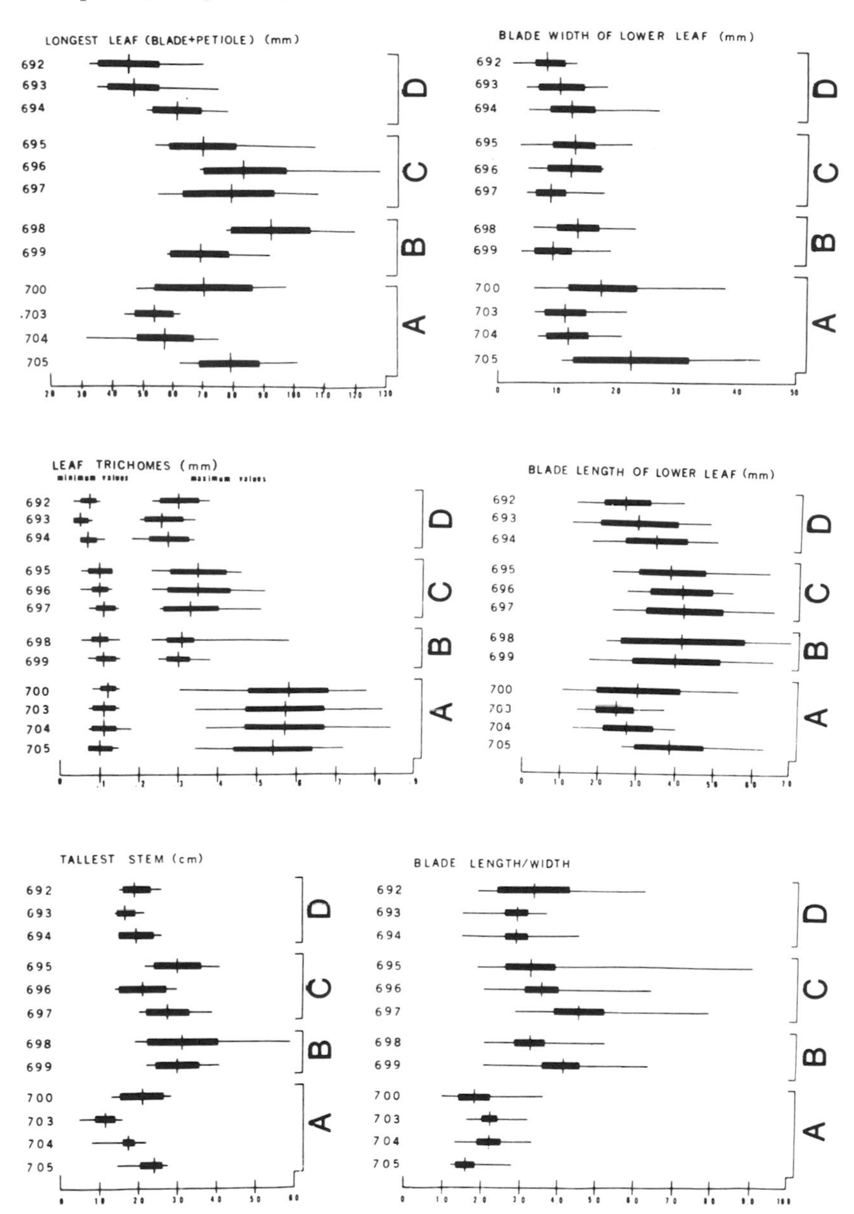

FIGURE 5.1.2 Quantitative measurements of some characteristics of taxa A–D. The range of variation observed is indicated by the length of the horizontal line. The resulting unweighted mean and twice the standard error are indicated by the vertical line and accompanying rectangle.

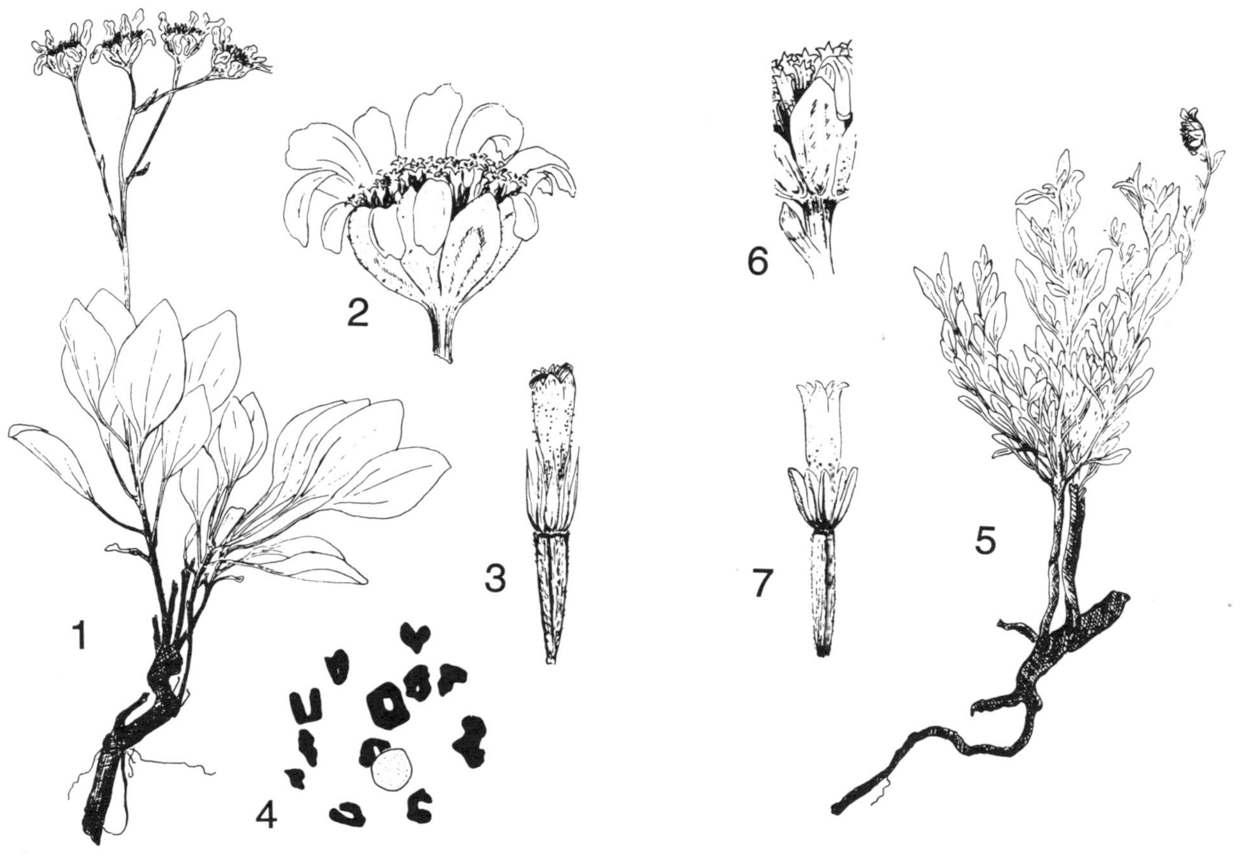

FIGURE 5.1.3 Drawings of features of taxa A (1–4) and D (5–7). Entire plant, 1 and 5, ×0.5; head, 2 and 6 (part only), ×3.5; disc floret, 3 and 7, ×5; chromosomes in diakinesis, 4, ×1000.

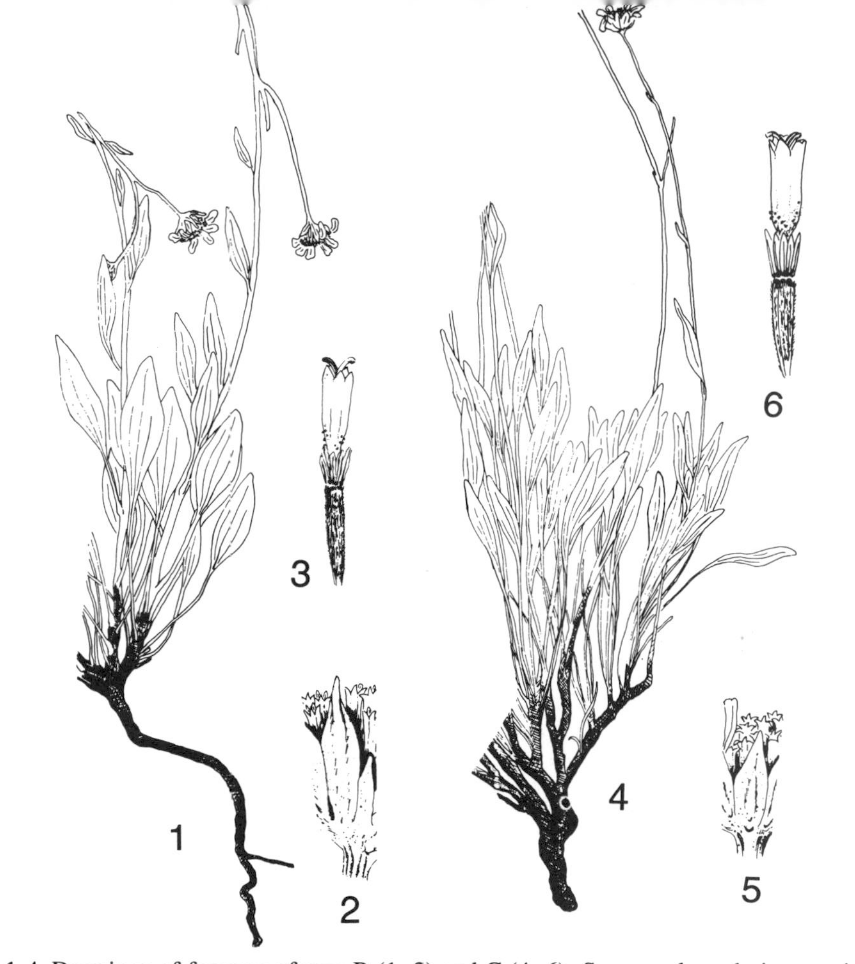

FIGURE 5.1.4 Drawings of features of taxa B (1–3) and C (4–6). Same scale and views as in figure 5.1.3.

TABLE 5.1.1.
Some Significant Morphological Characters of Taxa A–D

Character	Taxon A	Taxon B	Taxon C	Taxon D
Stem leafiness	mostly basal or occasionally extending midway up stem	both basal & cauline, leaves extending up stem 15–60% of stem height	both basal & cauline, leaves extending 20–60% of stem height	mostly leafy to apex or rarely only midway
Lower stem surface processes	glabrous to moderately pubescent w/ few–numerous stipitate glands	lightly to heavily pubescent w/ few globules of surface exudate	lightly to moderately pubescent w/ stipitate glandulosity increasing markedly w/ height	glabrous to heavily pubescent w/no–few short stipitate glands
Upper stem surface processes	moderately pubescent w/ few–numerous stipitate glands	lightly to heavily pubescent w/ no–few stipitate glands	glabrous to moderately pubescent w/ numerous stipitate glands	lightly to moderately pubescent w/ no–few sessile or short stipitate glands
Peduncles & bract surface processes	numerous short stipitate glands	no to moderate pubescence w/ no–few sessile or short stipitate glands; occasionally glabrate	numerous stipitate glands	lightly to heavily pubescent w/ few– numerous stipitate glands
Petiole length (mm)	(3)12–45	(7)13–40	16–60	8–35
Peduncle length (mm)	(2.5)8–55(65)	11–140	6–140	8–55(75)
Bract apex	obtuse, acute	caudate-attenuate or cuspidate (rarely acuminate)	obtuse, acute	obtuse, acute, acuminate
Number of disc florets	41–83	40–60	40–75	25–58
Disc floret throat length (mm)	2.3–3.8	1.4–3.5	1.5–4.0	2.0–2.8
Achene length (mm)	1.0–5.1	2.0–6.2	3.5–6.0(8.0)	3.0–4.6
Pappus scale length (mm)	(0.6)1.0–3.2	0.7–3.2	0.7–3.7	1.0–2.4(2.8)

DISCUSSION: The populations mapped in figure 5.1.1 are all the known collections (at the time this paper was published) of a small genus, *Platyschkuhria* Rydberg, in the sunflower family (Compositae). Four taxa of some level within the genus have been delimited. The problem, then, is to determine how many species or infraspecific taxa (if any) are involved.

The morphological differences among the four taxa, obviously, are very slight, and this suggests that all belong to the same species. A graph of quantitative features (figure 5.1.2) reveals gradations in all characters, except that Taxon A seems to stand out as somewhat distinct from the rest. Looking at the drawings of the plants in figures 5.1.3 and 5.1.4, we notice that Taxon A does seem to be the most different of the group, especially in the shape of the leaves and pappus scales. The distribution of the known populations (figure 5.1.1) shows those of Taxon A (with one exception) to be quite separated from the rest, perhaps implying ecological differentiation. The other three taxa (B, C, and D) seem closer morphologically and distributionally.

Based on these data alone, one might wish to place Taxon A in a subspecies by itself, with the other three taxa in a second subspecies that has three distinct varieties. The author of the original paper, however, treated all taxa as separate and coordinate varieties, with comments indicating that Taxon A was believed somewhat more different than the rest. A more recent study by Brown (1983) documented different chromosome (ploidy) levels within Taxon C (diploid, tetraploid, pentaploid, and hexaploid), which complicates the picture even more. Brown (1983), however, did not suggest altering the four varieties proposed by Ellison (1971). Brown also made more collections, and the southern outlier of Taxon A in south-central Utah was redetermined to be actually Taxon C.

Case Study Example 2 (Orchidaceae)

PROBLEM: Provide a classification for the two taxa (A and B) using any of the following categories: genus, species, subspecies, variety, and form.

MATERIALS, METHODS, AND SUPPLEMENTAL DATA: Taxon A, widespread in northeastern North America, grows in cool, wet, neutral, often highly organic soils, in habitats varying from the full sun of roadside ditches and wet meadows to the shade of seasonally wet woodlands. Taxon B is restricted to Appalachian uplands in the south but grows at progressively lower altitudes northward, reaching sea level in the New England states. It grows in soils similar to those characteristic of Taxon A habitats and also grows in full sun to shade, frequently under or near conifers. Taxon B is collected more often in well-drained soils of upland conifer or northern hardwood forests than is Taxon A. (See figure 5.2.1 for a comparison of their distributions and figure 5.2.2 for drawings of their key morphological features.)

The skipper, *Polites mystic,* and hawkmoth, *Haemorrhagis thysbe,* have both been observed visiting roadside colonies of Taxon A near Grand Marais, Michigan. The skipper alone has been observed visiting plants near East Cabot, Vermont. Skipper visitors withdraw the pollinarium in a position perpendicular to the shaft of their proboscis. The pollinarium is attached 1–2 mm from the head on the 13–15 mm long proboscis. Within 15 seconds of withdrawal the stalk and attached pollen masses begin to move forward to a position more nearly parallel to the proboscis. The movement is effected at the stalk-viscidium junction and is usually completed within two minutes; during this time the pollinia move through an angle of 25–60°. The proboscis length is less than the mean nectary length (see below). The hawkmoth also transports pollen with the pollinarium attached 3–5 mm from the head on the 20 mm long proboscis. Mean nectary length in Taxon A is uniform throughout the latitudinal range of the taxon, while mean nectary length of Taxon B is greater in northern races than in the central and southern Appalachian regions (figure 5.2.5). These variations in nectary length may indicate different pollinator species in various portions of the range of taxon B.

Taxon B has been observed in the field near East Cabot, Vermont, and Cranberry Glades, West Virginia, but without observation of lepidopteran visits. The oval viscidia are appropriately spaced for attachment to the compound eyes of medium- to large-sized lepidoptera. When a glass rod is touched to the viscidia and withdrawn, the stalks are perpendicular to the rod and parallel to each other. They then change position, converging toward each other and moving to the front of the viscidium attachment. This movement also originates at the stalk-viscidium junction. The massulae (divisions of the sectile pollen mass) are now in position to enter the stigmatic depression between the anther thecae. The forward motion was recorded at 15-second intervals for pollinaria of the two species, and there are differences in the apparent rate of angular movement but not in the time required for completion.

When herbarium collection dates are compared with latitude, the flowering periods of the taxa are seen to overlap (figure 5.2.3). The records span more than a century, and when annual climatic fluctuations are superimposed, they will to some extent obscure the actual sequence in a particular year. Figure 5.2.3 includes material widely separated longitudinally in the north, with no indication of phenological differentiation east to west. Taxon A flowers in early June in the southern Appalachian area and progressively later at higher latitudes. Taxon B has also been collected in flower during the first half of June in the Carolina–Virginia–West Virginia area. It tends to flower before most populations of Taxon A, but there is no clear separation. Taxon A has been observed in central New York state in bud at the same time that nearby Taxon B had been out of flower 10–14 days. Two days later in Vermont, colonies of the two taxa were flowering concurrently at sites three miles apart. Local climatic and edaphic conditions may strongly affect such flowering sequences and may be responsible in part for the observed differences.

Flowering sequences are more complex in Virginia and in North Carolina. There, Taxon B has been collected and observed to flower in early August, six weeks later than local populations of Taxon A (Carlyle Luer, Charles Wright, pers. comm.). Both taxa occur at 5000–6000 ft altitude, and this disjunct flowering sequence cannot be due to altitudinal differences. More than one phenological race of Taxon B may be present in the southern Appalachians, each being adapted to different broods of pollinators.

Root tip chromosomes were examined from two plants each of Taxon A from Grand Marais, Michigan, and Taxon B from Cranberry Glades, West Virginia. In all plants the diploid number was 42.

A small number of specimens examined during this study suggest that hybridization may have occurred between Taxon A and Taxon B. A collection from King's County, Nova Scotia, 26 July 1944 (*Lewis & Gorman s.n.*,

DAL) has an unusually wide inflorescence (like Taxon B) but a column more similar to Taxon A. As flower color is usually poorly preserved and as labels have few notes concerning floral characters, it is very difficult to determine specimens as old as these.

Experimental crosses of plants in the greenhouse have been done in order to determine whether such combinations produce seeds. The results are presented in figure 5.2.4. Seeds with apparently normal embryos were produced in out-crossed combinations, although the parental plants tested were partially self-incompatible. Seeds of all combinations have been surface-sterilized and placed on Knudson C agar in darkness, a procedure that is effective in germinating some terrestrial species of this family. Repeated attempts to germinate seeds of Taxon A and Taxon B failed; later a single embryo of Taxon A began to grow nine months after sowing. Absence of germination cannot be used as sufficient evidence for incompatibility because cultures often behave erratically in the timing of germination, and seeds in this series were too limited in number to permit multiple cultures. It can only be said that attempts to cross members of Taxon A and Taxon B have largely failed. Intermediates between Taxa A and B have been mentioned in floras, but little evidence exists for hybridization in terms of column morphology in the herbarium material examined. Such hybrids may be difficult to detect if one column type is dominant.

DATA: figures 5.2.1 through 5.2.5.

REFERENCE: Stoutamire, W. P. 1974. *Brittonia* 26: 42–58.

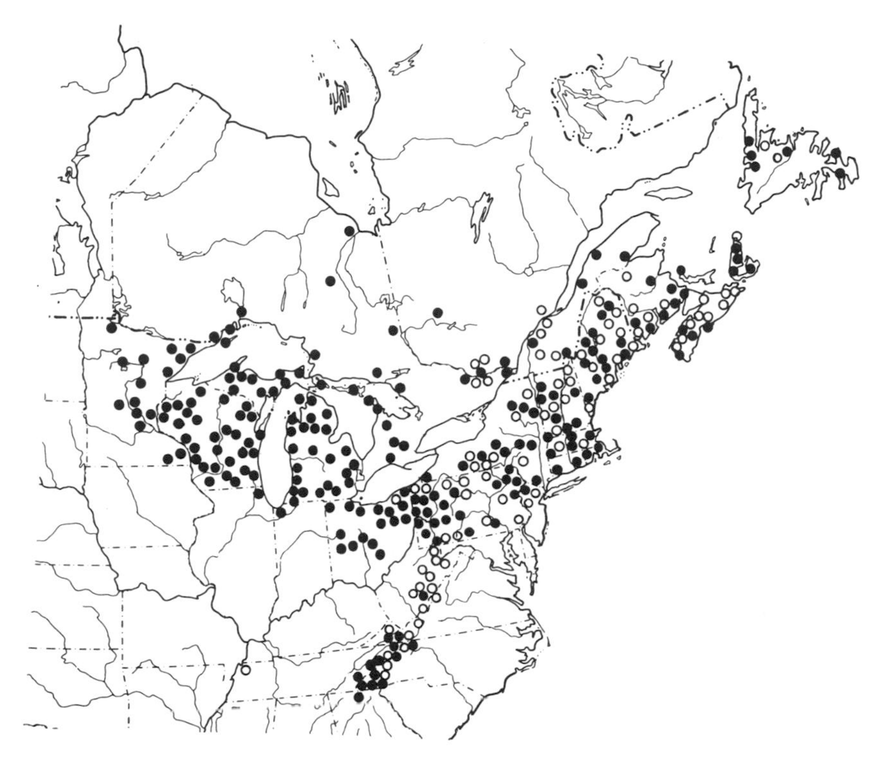

FIGURE 5.2.1 Map showing distribution of taxa A and B. Taxon A, blackened circles; Taxon B, circles.

FIGURE 5.2.2 Morphological features of taxa A (1–5; Grand Marais, Michigan) and B (6–10; Cranberry Glades, West Virginia). Drawings 1, 2, 6, and 7 are flowers (×2.5); 3 and 9 are pollinaria (×9); 4 and 8 are seeds (×25); 5 and 10 are habits (×0.2). Redrawn from photographs of living specimens.

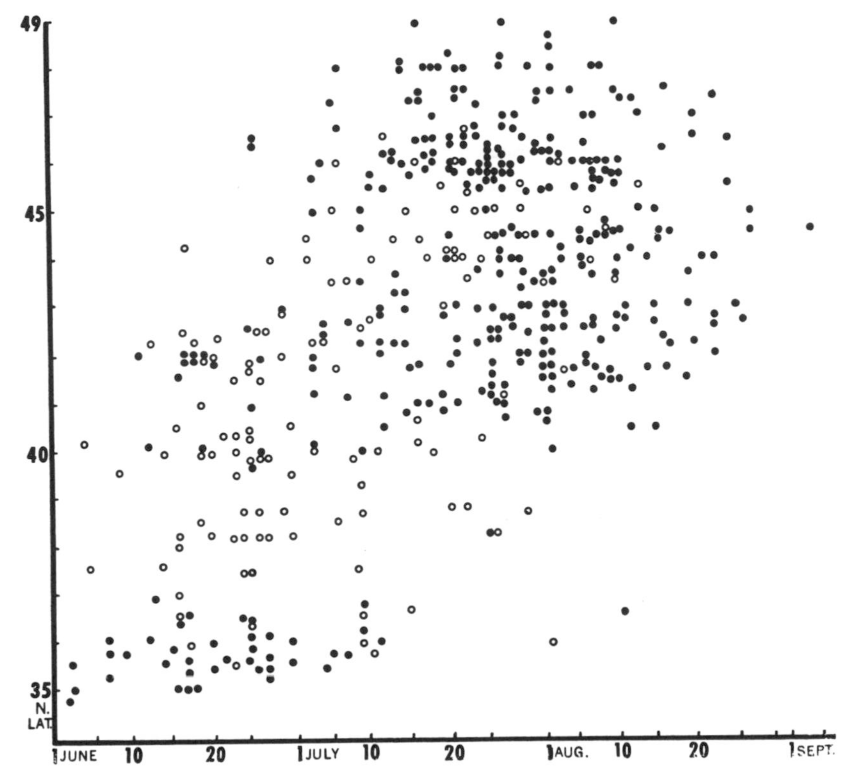

FIGURE 5.2.3 Correlation of latitude and collection dates of flowering specimens of Taxon A (blackened circles) and Taxon B (circles).

TAXON B
self 5/5, 2.5%

20/21, 42

28/31, 42

TAXON A
19/20, 16

FIGURE 5.2.4 Experimental crossing relationships between Taxon B (Cranberry Glades, West Virginia) and Taxon A (Grand Marais, Michigan). Numbers beneath taxa are selfing results; numbers with arrows are crossing results (arrows indicate direction of pollen transfer). The fractions indicate number of mature ovaries to the number of flowers pollinated; the percentage figures show the percentage of seeds with mature embryos.

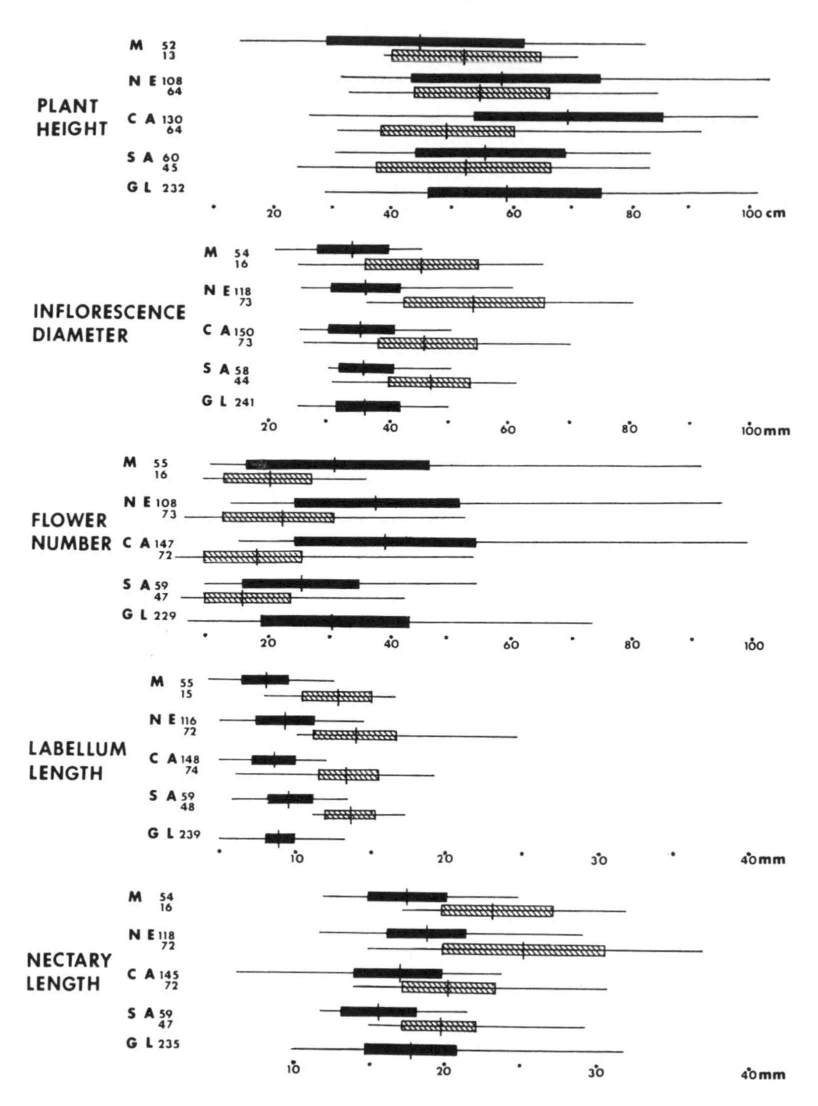

FIGURE 5.2.5 Ranges, means, and standard deviations of vegetative and floral characters in Taxon A (solid bars) and Taxon B (hatched bars) by geographic area. Numbers following area designation are sample size. Geographic areas are: M (Maritime) = New Brunswick, Nova Scotia, Newfoundland, Quebec; NE (Northeastern) = Maine, New Hampshire, Vermont, Massachusetts, Connecticut, Rhode Island; CA (Central Appalachian) = New York, Pennsylvania, New Jersey, Ontario; SA (Southern Appalachian) = West Virginia, Virginia, North Carolina, Tennessee, Georgia; GL (Great Lakes) = Ohio, Indiana, Michigan, Wisconsin, Minnesota.

DISCUSSION: This problem deals with taxa in Orchidaceae, which are noted for morphological adaptations to insect pollination. The genus in question is *Platanthera* Rich.

In assessing the relationships between Taxon A and B, it is useful to begin by looking at the morphological data. The two taxa obviously differ from each other in several floral characteristics. The most conspicuous differences are in the size and shape of the petals, the size and shape of the column (a fleshy structure containing stamen and carpels fused together), and length of stalk of the pollinia (i.e., of the pollen masses). In addition to these differences in floral features, the leaves show some slight differences in size. The data regarding these and other characters, plus a geographic correlation, are shown in more detail in figure 5.2.5.

Turning next to the distributional (figure 5.2.1) and ecological data, we notice that the two taxa overlap for most of their ranges, but tend to occur in different habitats. The flowering dates of the two coincide to a large degree (figure 5.2.3). Focusing lastly on the experimental crossing results, we learn that the two taxa are reasonably compatible. In fact, they cross with as much ease as plants within the same taxon; however, no hybrid progeny were successfully raised (see the narrative in supplemental data section).

Having reviewed all the data, we must conclude that the two taxa are very similar in many ways. At the same time, they do differ in several obvious respects. In my opinion, it is initially clear that Taxa A and B should be recognized as distinct at the varietal, subspecific, or specific rank. A decision as to which rank to use in this case depends largely upon the concepts of categories held by the worker, which is in part determined by an understanding of the historical approach to circumscribing taxa in this particular family (Orchidaceae). The author (Stoutamire) opts for specific recognition for both Taxon B and A, as *P. grandiflora* (Bigel.) Lindley and *P. psycodes* (L.) Lindley, respectively.

Stoutamire recommends species designation primarily because of the different pollinating agents, and therefore the probable absence of gene flow between the two taxa. He concludes: "I will treat the taxa as distinct species rather than as subspecies or varieties because they appear to be seasonally, mechanically, and spatially isolated, morphologically distinguishable, but badly confused entities" (p. 43). A similar conclusion was reached in the same genus by Sheviak and Bowles (1986) for another closely related species pair, *P. leucophaea* (Nutt.), Lindley and *P. praeclara* Sheviak & Bowles.

Case Study Example 3 (Compositae)

PROBLEM: Provide a classification for the three taxa (A, B, C) using any or all of the following categories: species, subspecies, variety, form.

MATERIALS AND METHODS: The taxa under study are spring-flowering shrubs in which the vegetative growth of one season will bear involucres the next, followed by a dieback of stems to the ground. All three taxa are characterized by segregation of staminate and pistillate florets into separate flowering heads; the pistillate florets are enclosed in woody, bur-like involucres with connate phyllaries that are usually specialized for dissemination. The wing-like processes (scales) of the involucres of the taxa under study are evidently modified for wind dispersal. Within this group there is considerable variation in pistillate involucre morphology. Some plants have the wings spirally arranged over and appressed to the involucre body, and others have the wings whorled and oriented at approximately right angles to the involucre axis.

For the morphological analyses, 136 specimens were selected for study. Selection was based on suitability of stages for comparing all characters and on representation for the entire geographic range (figure 5.3.1). Developmental stages were particularly important because the plants lose their staminate heads as pistillate heads develop. The problem is further complicated by the fact that specimens with mature fruits tend to be collected more often because the fruits are more visible. Of the 136 specimens studied, 50 had wings appressed and spirally arranged, 50 had wings whorled and emergent, and 36 had an intermediate arrangement. All measurements and observations reported in the accompanying tables and graphs were taken from this sample.

For microscopic examination, florets, pales, and staminate involucres were moistened, dissected, stained with 1% Bismarck brown in 70% ethanol, and mounted in Hoyer's mounting medium (Galigher and Kozloff. 1971. *Essentials of Practical Microtechnique*. Ed. 2. Philadelphia: Lea and Febiger). Leaves, which were cleared using the hot NaOH technique developed by Payne (1969. *Ward's Bull.* 61: 4–5), were stained with 1% safranin in 100% ethanol and with 1% Bismarck brown in 70% ethanol. Pollen stainability was evaluated by staining in acetocarmine.

DATA: figures 5.3.1 through 5.3.3; tables 5.3.1 and 5.3.2.

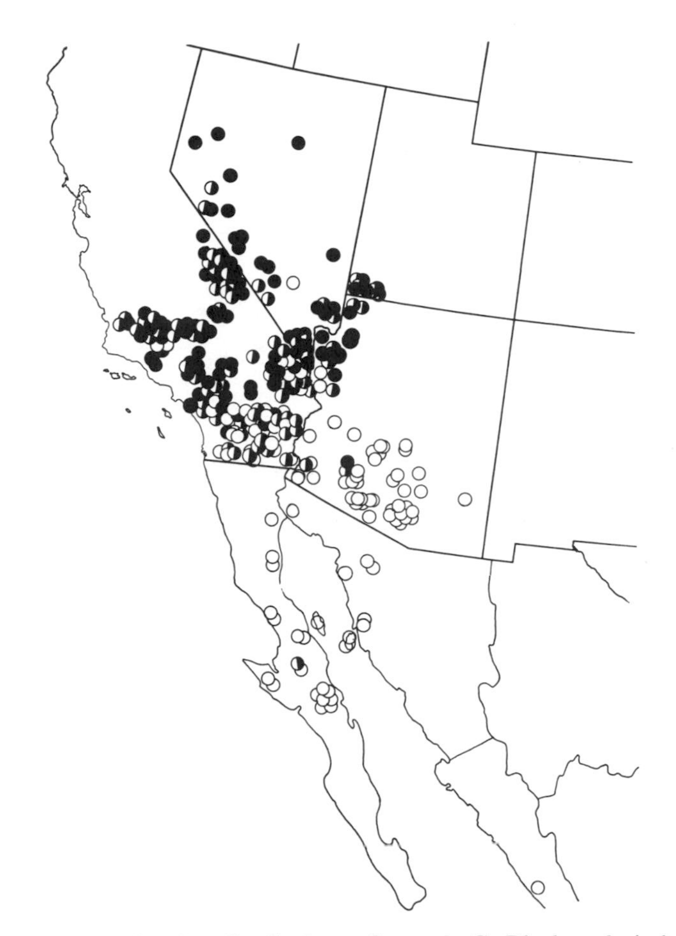

FIGURE 5.3.1 Map showing distributions of taxa A–C. Blackened circles represent Taxon A, circles represent Taxon C, and half-blackened circles represent Taxon B.

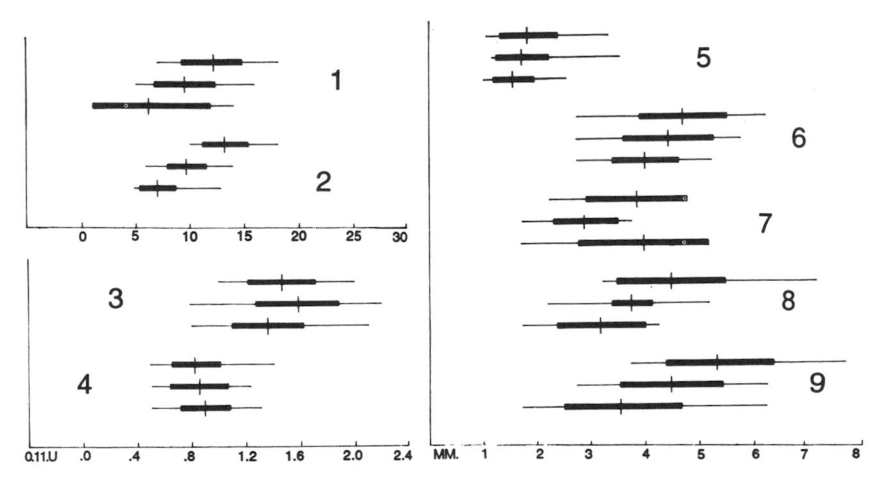

FIGURE 5.3.2 Ranges, means, and standard deviations of characters of taxa A–C (A, top; B, middle; C, bottom). Character 1, number of staminate florets per involucre; 2, number of wings per fruiting involucre; 3, length/width ratios of fruits; 4, length/width ratios of wings; 5, beak length; 6, fruit length; 7, fruit width; 8, wing length; 9, wing width.

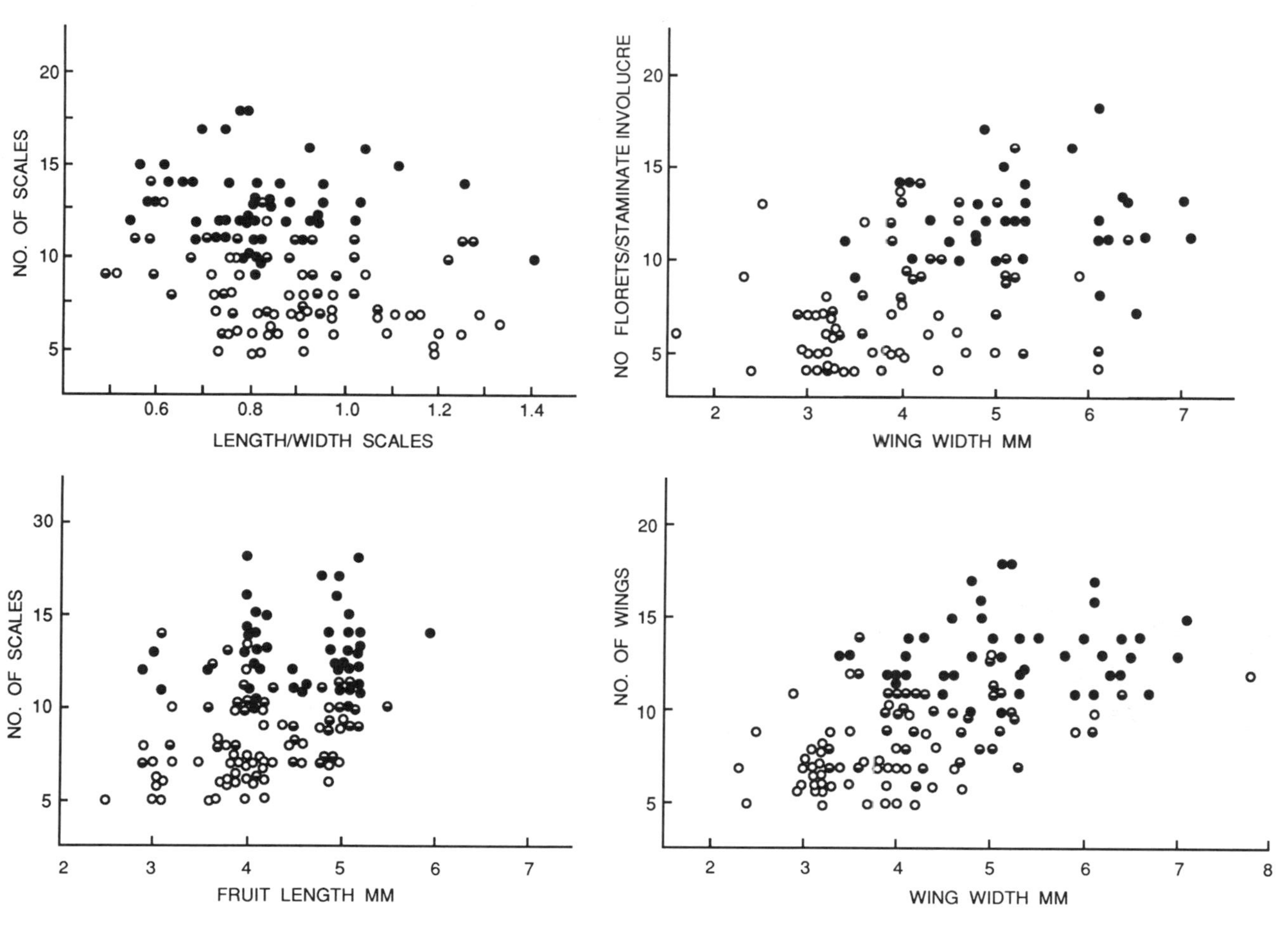

FIGURE 5.3.3 Scatter diagrams of best correlated characters of taxa A–C. Blackened circles from sample of Taxon A, circles from Taxon C, and half-blackened circles from Taxon B.

TABLE 5.3.1.
Data (in mm) from Taxonomically Correlated Characters in Taxa A–C

	Taxon A		Taxon B		Taxon C	
	Mean	S.D.	Mean	S.D.	Mean	S.D.
Wing Number	13.20	±1.98	9.70	±1.82	7.00	±1.77
Wing Length	4.51	1.01	3.79	0.37	3.20	0.82
Wing Width	5.38	0.98	4.50	0.94	3.59	1.07
Length/Width	0.82	0.18	0.84	0.21	0.89	0.19
Fruit Length	4.72	0.80	4.44	0.84	4.03	0.61
Fruit Width	3.88	0.92	2.90	0.58	4.00	1.21
Length/Width	1.46	0.24	1.58	0.30	1.36	0.26
Beak Length	1.82	0.54	1.75	0.52	1.60	0.39
No. Staminate Florets	12.12	2.77	9.64	2.81	6.25	5.65

TABLE 5.3.2.
Pollen Stainability in Taxa A–C

	Taxon A	Taxon B	Taxon C
Mean	72.2%	62.5%	71.0%
Range	54–87%	35–93%	43–96%

REFERENCE: Peterson, K. M. and W. W. Payne. 1973. *Brittonia* 25: 243–256.

DISCUSSION: This problem deals with the morphologically somewhat unusual genus *Hymenoclea* Torrey & A. Gray, of the Compositae. The case study treats the spring-flowering complex of populations in the southwestern United States and adjacent Mexico. For a helpful analysis of this problem, the authors themselves provide the best discussion:

"Specimens of the appressed form [Taxon A] typically have wings of the fruiting involucres scattered over (spirally arranged on) the involucre body . . . , and those of the whorled form [Taxon C] have wings disposed more or less in a single, central whorl. . . . Every shade of intergradation of wing placement and of appression exists between these two extremes [Taxon B]. Of the other characteristics examined, wing number and wing size showed the best taxonomic correlations. In general, specimens of the appressed form have more, larger wings, and those of the whorled form have fewer, smaller wings. . . . Also, there is a general trend in fruit size and beak length from longer in the appressed form to shorter in the whorled form. Again, however, there is broad overlap between the ranges of all these character expressions; when these best-correlated features are plotted as scatter diagrams [figure 5.3.3], we do not find taxonomic clustering. Thus, in all characteristics examined, there is marked intergradation and overlap in expression from one group to the other, suggesting continuous variation incompatible with the usual nature of distinct species'' (pp. 248, 250).

''To further test the hypothesis of distinct species, pollen stainability was examined with acetocarmine staining. The results, obtained from 10-specimen subsamples randomly selected from within each subgroup, are presented in table [5.3.2]. The ranges and means are very similar and are not largely depressed in the intermediate group as expected for interspecific hybrids. . . . Furthermore, plants in the intermediate . . . group showed no discernible diminution of seed or fruit production. Thus, there is no convincing evidence from indirect assessment of reproductive biology that the extremes represent different species'' (p. 250).

''The situation is brought into focus dramatically by considering the geographic distribution of the complex. In [Figure 5.3.1] all available specimens, assessed on the basis of fruiting involucre morphology as belonging either to one of the two morphological extremes or to the intermediate [taxon], are plotted geographically. Here we see a picture of two geographically oriented subpopulations, one northerly, one southerly, that are united by a broad area of overlap in which intermediate expressions abound. The geographic picture, in which there is widespread intermediacy and intergradation between two extremes, mirrors the situation found in analyzing characters independently'' (p. 250).

''Thus, this complex presents a near classic picture of one species with two derivative varieties, occurring in different climatic areas and with different adaptive peaks, overlapping in a broad area of ancestral intermediacy and of morphological intergradation. It is taxonomically helpful to recognize these as three intergrading varieties within the species *H. salsola.*'' (p. 253).

Although the authors' taxonomic conclusion of recognizing three separate varieties is one that can be suggested by the data, an alternative approach would be to recognize only two varieties (Taxa A and C), with the populations of Taxon B regarded as morphological intermediates or hybrid types. The slightly lowered pollen stainability of Taxon B gives some support for this view.

6. Case Study Exercises

The case study exercises are presented in essentially the same fashion as the case study examples. The concluding discussion sections, however, have been omitted, because now the reader is obliged to work out his or her own classification and the reasons substantiating it. As in the case study examples, the data have been taken from actual papers as indicated by the reference cited (again, deliberately without title). The exercises involve different types and presentations of data commonly employed in plant taxonomic studies. After working out the classifications, readers are encouraged to check on the original author's recognition of taxa and resultant ranking. (More details of recommended classroom procedure are given in the Preface.)

Case Study Exercise 1 (Gramineae)

PROBLEM: Provide a classification of the two taxa (A and B), using any of the following categories: species, subspecies, variety, form.

MATERIALS AND METHODS: The external morphology of the study group was examined both in the field and on herbarium specimens. A complete set of voucher specimens has been deposited in US and duplicates have been distributed to various herbaria.

For the anatomical studies field-collected leaf blades from the mid-culm region were fixed in FAA. After dehydration in an increasing ethanol series and infiltration with xylene, the leaves were embedded in paraffin, sectioned at 10 μm thickness, and stained with safranin/fast green (Berlyn and Miksche. 1976. *Botanical Microtechnique and Cytochemistry.* Ames: Iowa State University Press). Samples were examined and photographed on an Olympus BH-2 photomicroscope using Kodak Technical Pan film. Anatomical descriptions were completed following the procedure for standardizing comparative leaf anatomy as outlined by Ellis (1976. *Bothalia* 12: 65–109). For purposes of comparison and standardization, primary, I° vascular bundles (first order) are defined as those containing large metaxylem vessels on either side of the protoxylem elements, and they are usually associated with sclerenchyma girders or strands (Ellis 1976). All other vascular bundles are considered as tertiary, III° (third order), and contain indistinguishable xylem and phloem elements.

Scanning electron microscopy of the leaf surface was conducted by direct mounting of the blade on an aluminum stub and coating the specimen with approximately 15 nm of gold palladium. Samples were examined using a Cambridge S-100 scanning electron microscope and photographed with Polaroid 55 P/N film. Terminology for epidermal features follows Ellis (1979. *Bothalia* 12: 641–671) and, in part, Metcalfe (1960. *Anatomy of the Monocotyledons. I: Gramineae.* Oxford: Clarendon Press).

Floral buds were field collected and fixed in ethanol-acetic acid (3:1, V:V) prior to storage under refrigeration in 70% ethanol. Meiotic chromosome counts were obtained from acetocarmine squashes of pollen mother cells. Representative cells were recorded with sketches and photographed using Kodak Technical Pan film. Chromosome number determinations were based on observations of 15 or more cells from at least three individuals per population.

DATA: figures 6.1.1 through 6.1.4; table 6.1.1.

REFERENCE: Peterson, P. M. and C. R. Annable. 1990. *Syst. Bot.* 15: 515–525.

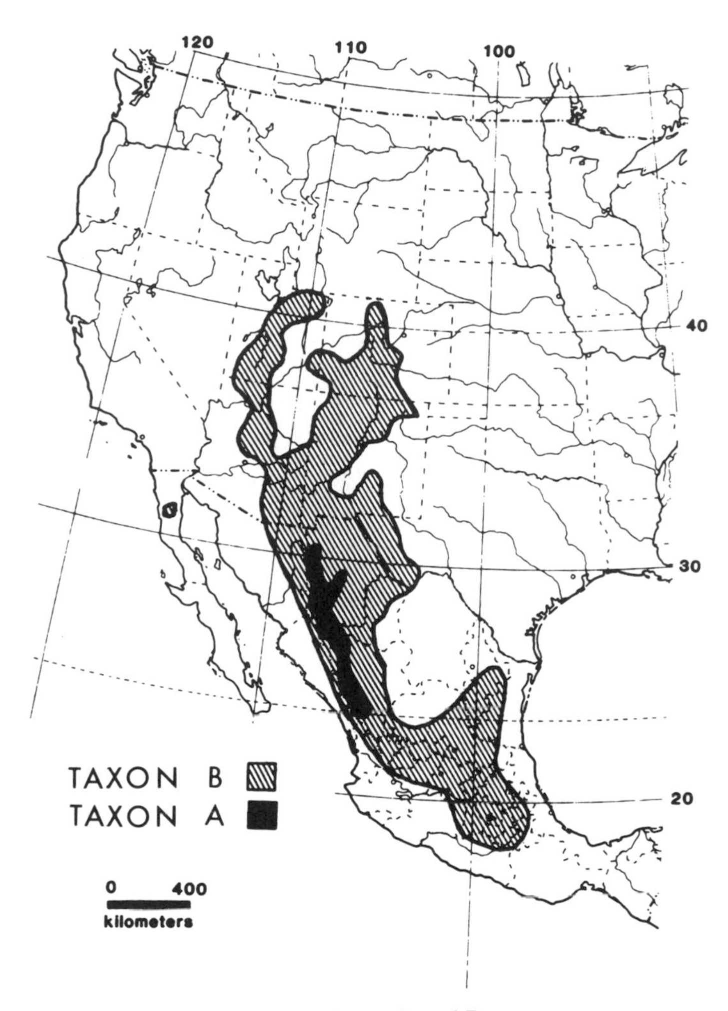

FIGURE 6.1.1 Geographic ranges of taxa A and B.

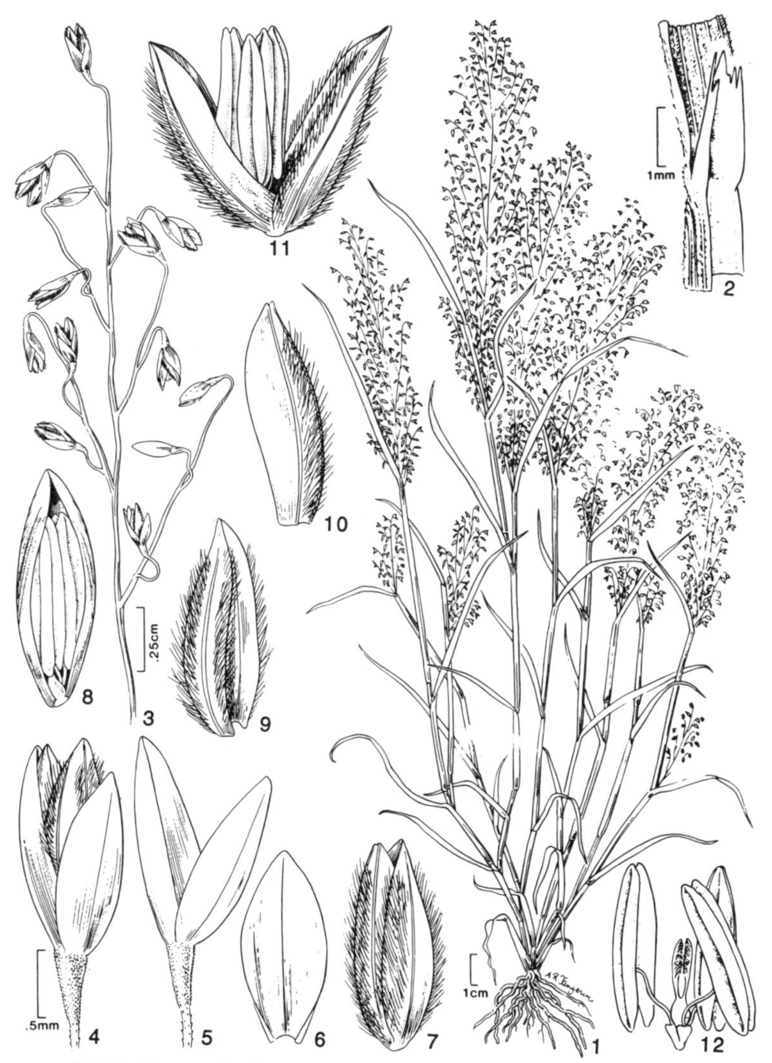

FIGURE 6.1.2 Taxon A, Durango, Mexico (*Peterson & Annable 4567*). 1, habit; 2, ligule; 3, inflorescence; 4, spikelet; 5, glumes; 6, lower glume; 7, floret; 8, stamens inside of palea; 9, lemma; 10, palea; 11, floret with stamens; 12, stamens, pistil, and lodicules.

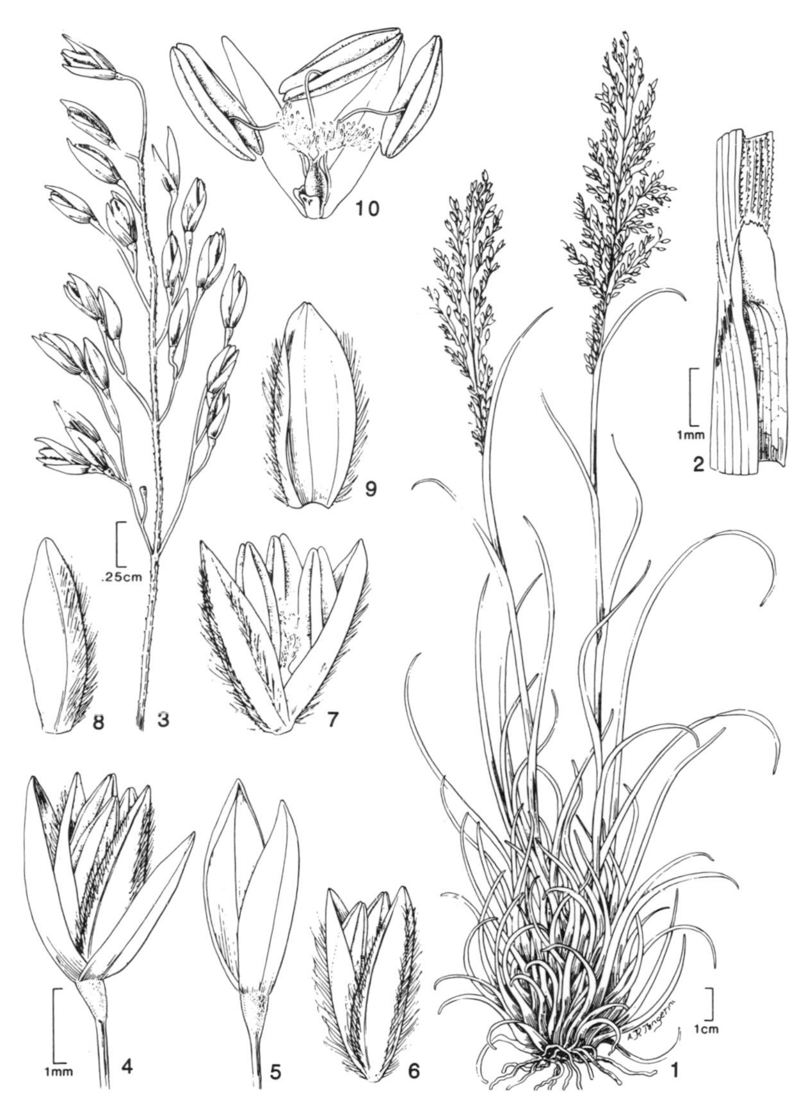

FIGURE 6.1.3 Taxon B, Grant Co., New Mexico, U.S.A. (*Peterson & Annable 5567*). 1, habit; 2, ligule; 3, inflorescences; 4, spikelet; 5, glumes; 6 & 7, florets; 8, palea; 9, lemma; 10, stamens, pistil and lodicules.

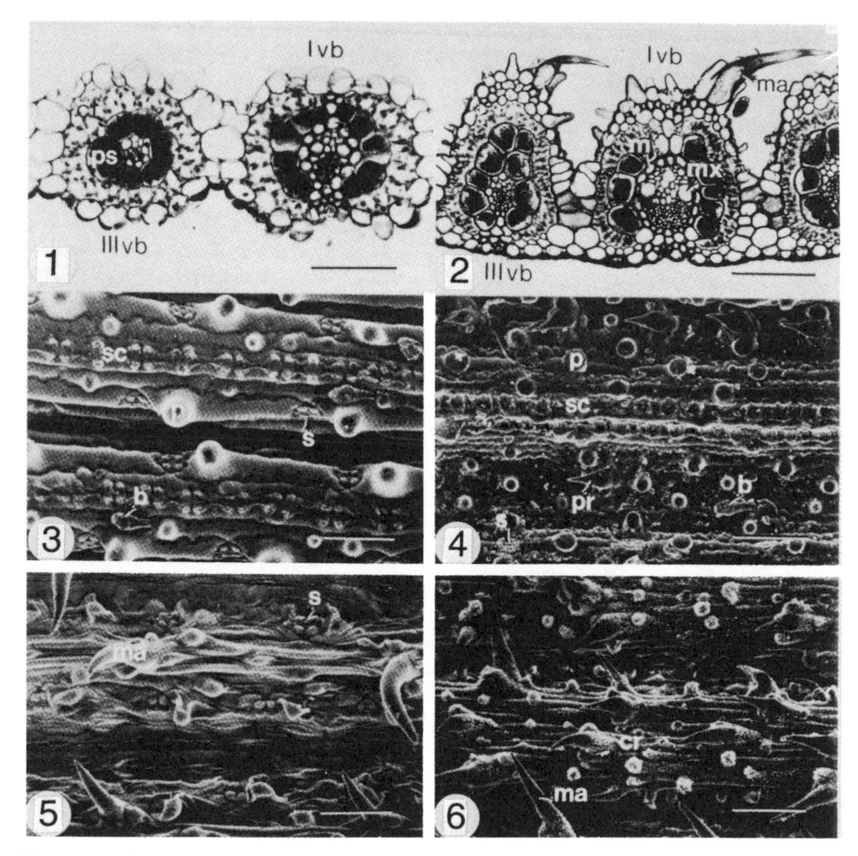

FIGURE 6.1.4 Leaf anatomy in taxa A and B (adaxial surface uppermost in figures 1 and 2). 1, Taxon A, with abaxial and adaxial ribs resembling a string of beads in cross section. 2, Taxon B, without prominent adaxial ribs. 3, Taxon A in abaxial view with prominent ribs and numerous dome-shaped stomata. 4, Taxon B in abaxial view showing a relatively flat surface and numerous prickle hairs. 5, Taxon A in adaxial view with numerous dome-shaped stomata and no crystaloid epicuticular projections. 6, Taxon B, in adaxial view with crystaloid epicuticular projections. Scales = 50 μm. b = bicellular microhair; cl = chlorenchyma; cr = crystaloid epicuticular projections; m = mestome sheath; ma = macrohair; mx = metaxylem element; p = papillae; pr = prickle hair; ps = parenchyma bundle sheath; s = stomata; sc = silica/cork cell; Ivb = primary vascular bundle; IIIvb = tertiary vascular bundle.

TABLE 6.1.1.
Chromosome Numbers (n) in Taxa A and B

Species	$n=$	Collector & number	Locality	Reference
Taxon A	8	*P & A 4122*	MEX. Durango: El Salto	Peterson 1988
	8	*P & A 4561*	MEX. Durango: El Ojito	Peterson 1988
Taxon B	8	*R & R 3133*	MEX. Durango: El Salto	Reeder 1971
	8	*R & R 4678*	MEX. Durango: Navios	Reeder 1971
	8	*R & R 5457*	USA. Arizona: Grand Canyon	Reeder 1977
	8	*R & R 5501*	USA. Arizona: Apache Co.	Reeder 1977
	8	*R & R 5507*	USA. Arizona: Apache Co.	Reeder 1977
	8	*R & R 5541*	USA. Arizona: Cochise Co.	Reeder 1977
	8	*Ward 86-058*	USA. Arizona: Cochise Co.	NMC herb. sheet
	8	*P & A 5457*	USA. Arizona: Cochise Co.	This paper
	8	*R & R 5145*	USA. Colorado: El Paso Co.	Reeder 1977
	8	*R & R 5430*	USA. Colorado: Ouray Co.	Reeder 1977
	8	*R & R 5434*	USA. Colorado: La Plata Co.	Reeder 1977
	8	*R & R 5888*	USA. New Mexico: Colfax Co.	Reeder 1977

NOTE: Collector abbreviations: *R & R* = Reeder & Reeder; *P & A* = Peterson & Annable.

Case Study Exercise 2 (Fagaceae)

PROBLEM: Provide a classification of the 20 taxa (A–T) belonging to one genus into two or more subgenera.

MATERIALS AND METHODS: Pollen samples were collected from specimens from several herbaria (with permission from the curators). Anthers were soaked for 24 hr at 60° C in 3% trisodium phosphate (Benninghoff. 1947. *Science* 106: 325)) or for 5 min at 95° C in 5% potassium hydroxide, acetolyzed (Faegri and Iverson. 1975. *Textbook of Pollen Analysis*. New York: Hafner Press), immersed in methyl alcohol, and filtered through 13-mm-diameter membrane filters, with 5-μm pore size. The filters with adherent pollen were folded to form closed envelopes, immersed in a 1:4 acetone-water solution, and dehydrated for 24 hr in a pure acetone atmosphere in the presence of anhydrous calcium chloride and anhydrous cupric sulfate. Each filter was dried by the critical point method (Polaron Model E-3000) with liquid carbon dioxide, then unfolded, pressed onto, and finally peeled off an SEM mounting stub which had been lightly coated with liquid adhesive (Dow Corning 289). The pollen that adhered to the mounting stubs was sputter-coated (Polaron Model E-5100 Series II) with gold, and examined in an SEM (International Scientific Instruments Model 1S1-40, maximum resolution 70 nm [Å]).

Samples were not included in the subsequent analysis if (1) fewer than ten pollen grains appeared in a single traverse of the 15-mm mounting stub, (2) debris adhered to pollen and obscured surface features, or (3) recognizable contaminant pollen appeared in the sample. Preliminary inspection eliminated 69 samples, leaving 102 samples. Detailed examination involved inspection of at least 20 pollen grains in each sample at 15k× (k× = thousand magnifications) or more. Experience indicated that no new diagnostic morphological characters appeared at magnifications higher than 10k×.

At least two representative grains from each sample were photographed on split screen at 2k× (whole grain) and 10k× (equatorial intercolpal area). More than two pollen grains were photographed in samples which contained pollen of multiple morphological types.

DATA: figures 6.2.1 through 6.2.5.

REFERENCE: Solomon, A. M. 1983. *Amer. J. Bot.* 70: 481–507.

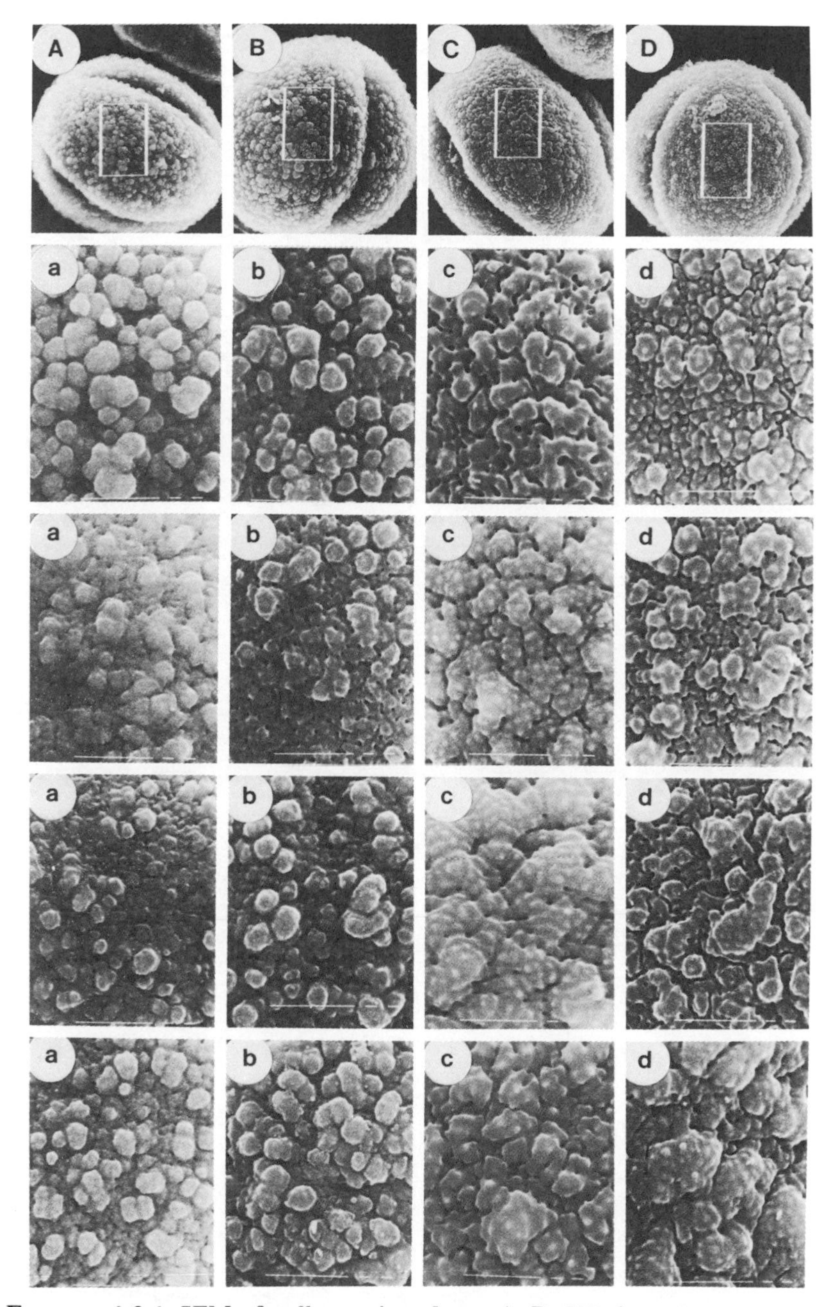

FIGURE 6.2.1 SEM of pollen grains of taxa A–D. Whole grains in top row. The four views below each whole grain (lower case letters) are intercolpal regions from grains of four different specimens. Whole grains = ×1,250; enlargements ×6,250. Bar = 2 μm.

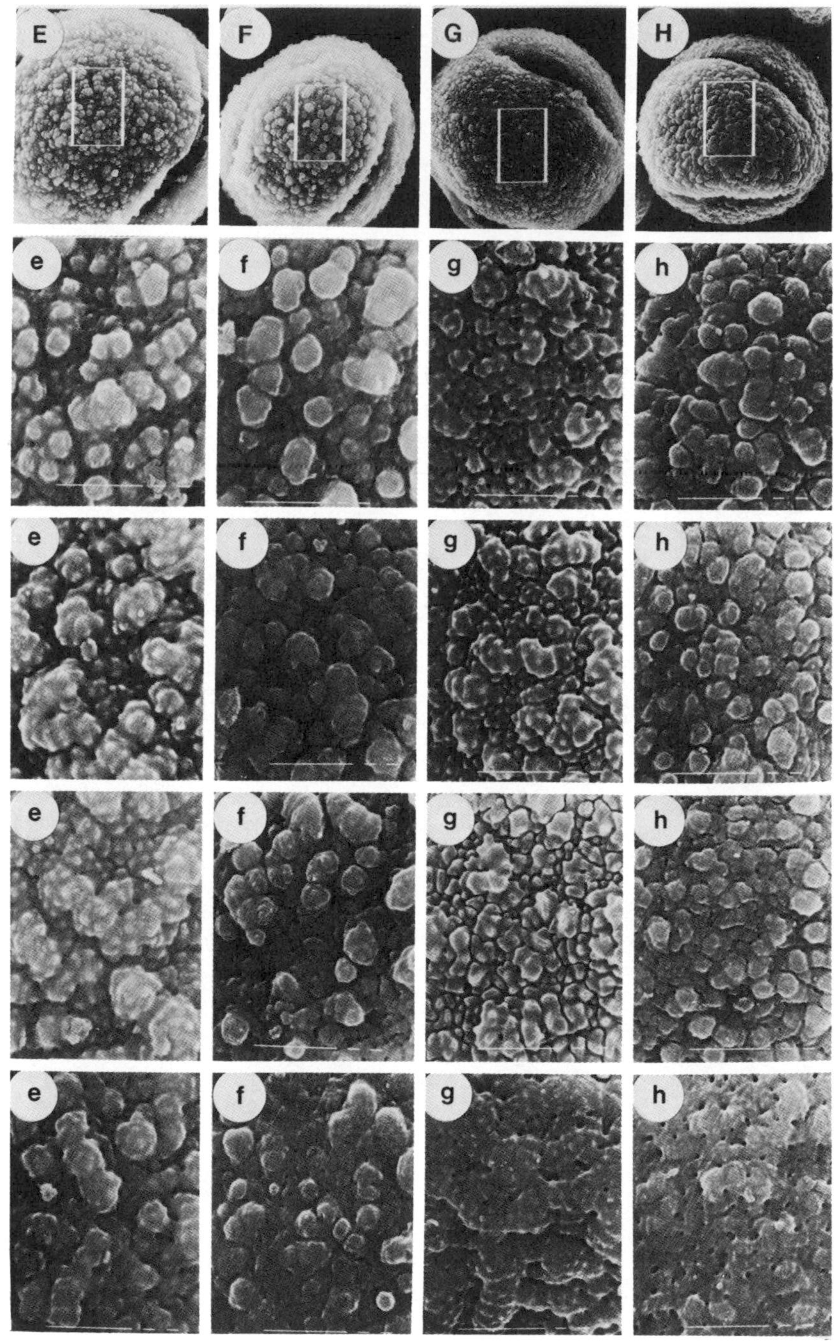

FIGURE 6.2.2 SEM of pollen grains of taxa E–H. Views and scale as in figure 6.2.1.

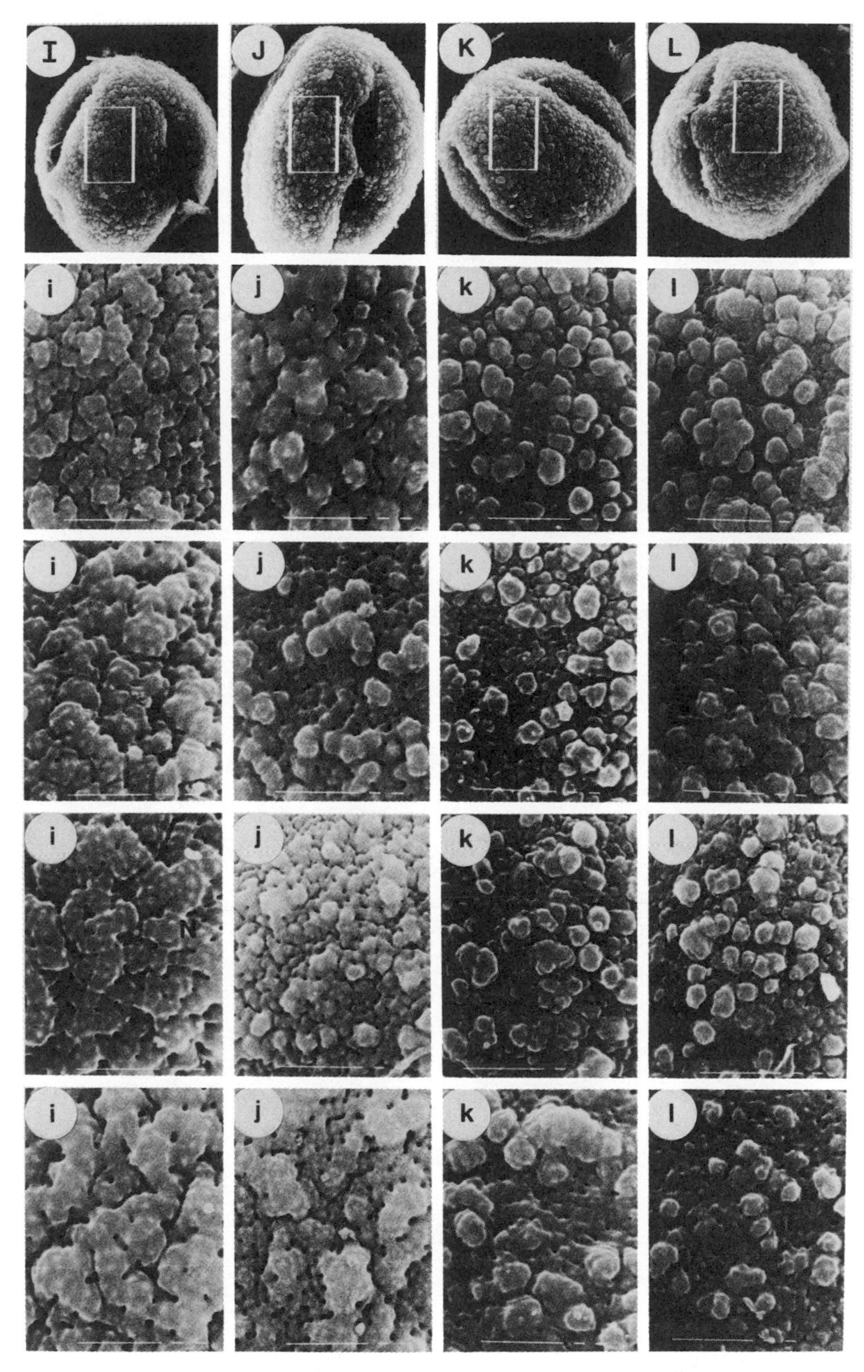

FIGURE 6.2.3 SEM of pollen grains of taxa I–L. Views and scale as in figure 6.2.1.

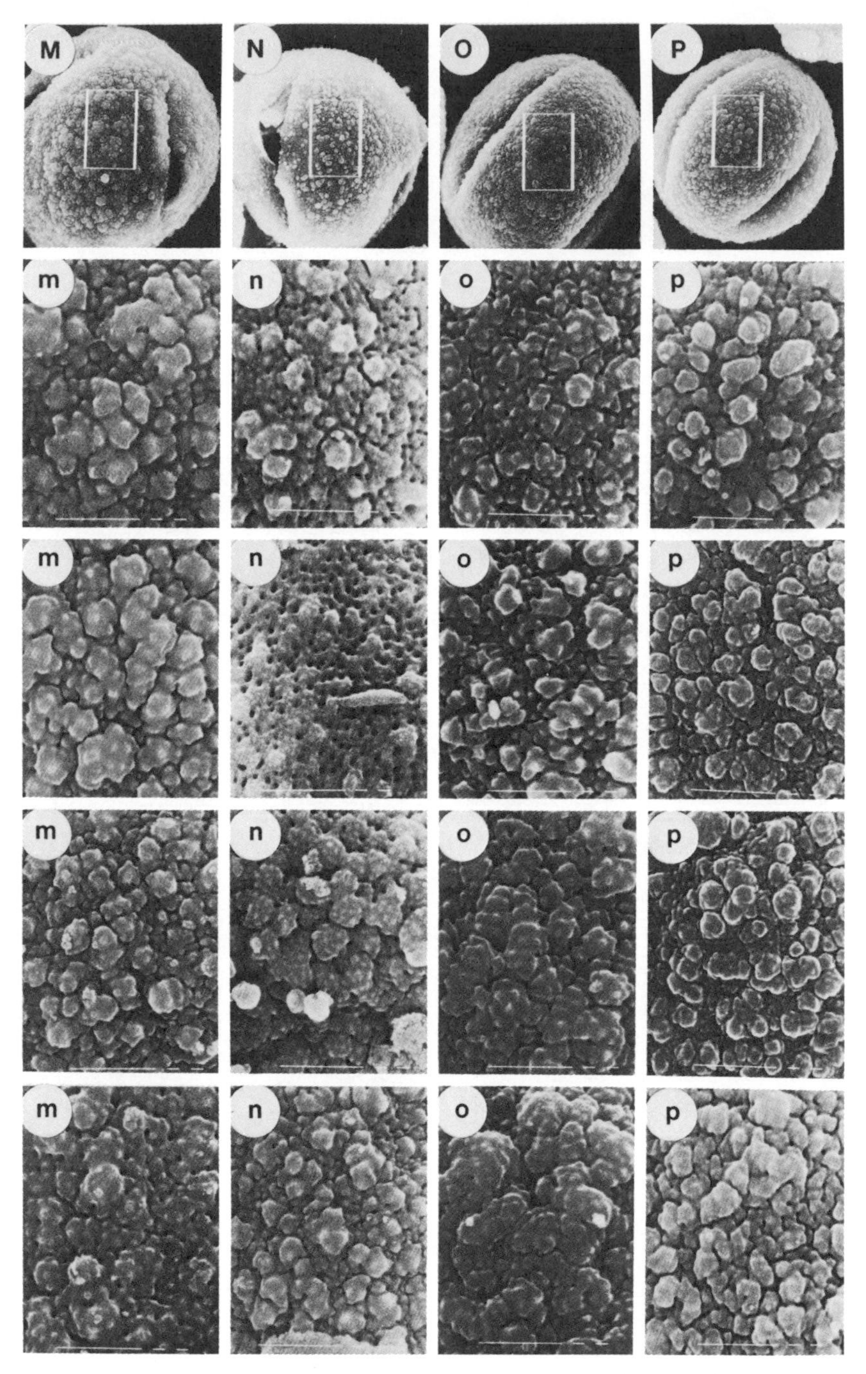

FIGURE 6.2.4 SEM of pollen grains of taxa M–P. Views and scale as in figure 6.2.1.

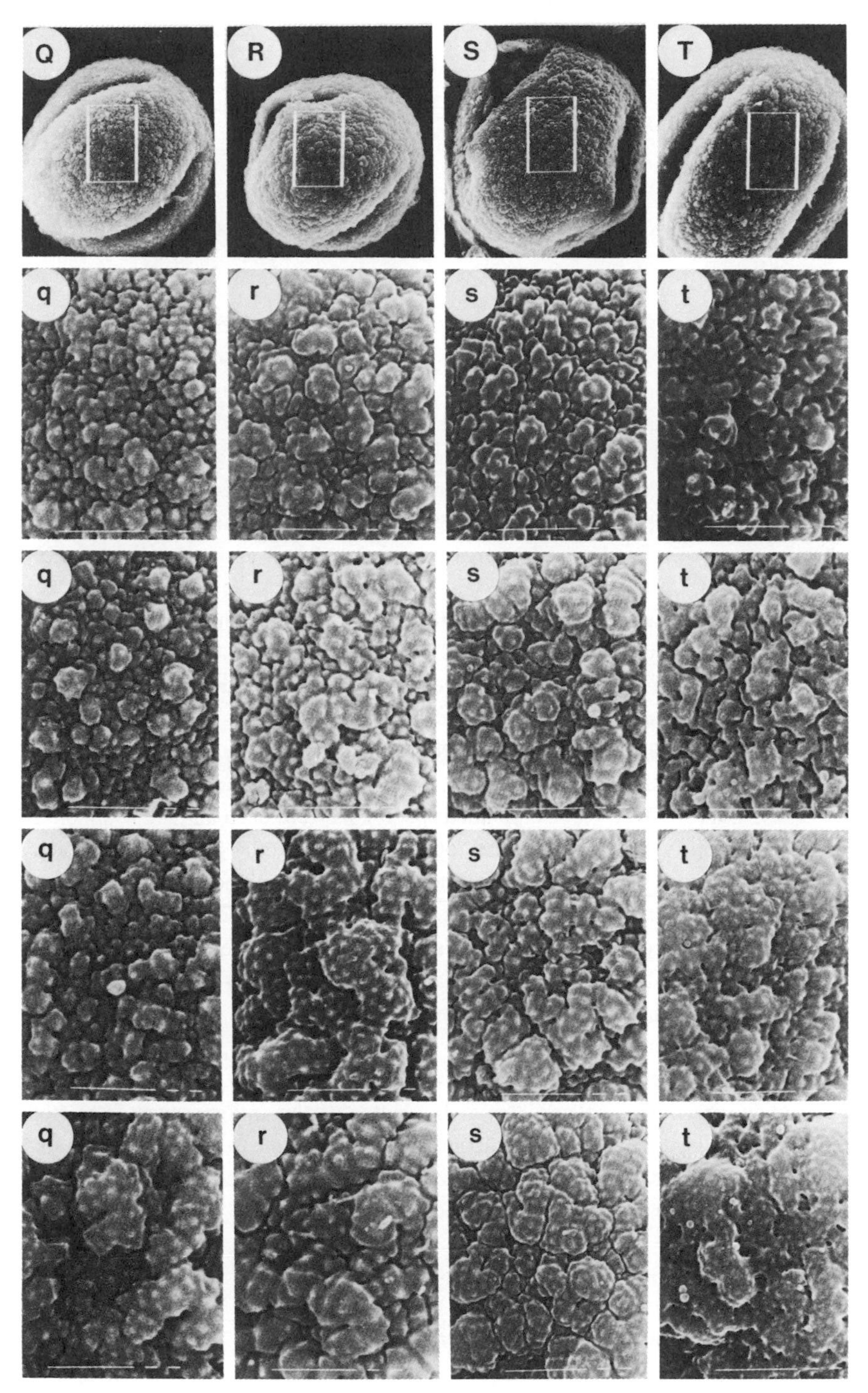

FIGURE 6.2.5 SEM of pollen grains of taxa Q–T. Views and scale as in figure 6.2.1.

Case Study Exercise 3 (Compositae)

PROBLEM: Provide a classification of the 28 populations included in this study, using any or all of the following categories: species, subspecies, variety, form.

MATERIALS AND METHODS: Twenty-eight populations (see figure 6.3.1) were sampled by collecting 25 specimens from each population, with each specimen collected from a different clump. From the population samples the following characters were measured or scored: length of leaf blade; width of leaf blade; leaf blade length/width ratio; involucre length; involucre width; involucre length/width ratio; width of widest involucral bract; shape of bract tip [distinguished as subacute to rounded (1), acute to acuminate (2)]; number of flowers per head; and leaf margin [distinguished as entire (1), remotely serrate (2), denticulate (3)]. These characteristics have been used as key characters in past treatments, or were found from preliminary studies to have possible taxonomic value.

Herbarium material was borrowed from the major national herbaria and from the regional herbaria within the range of the taxa (A, COLO, DUL, FLAS, FSU, GA, GH, IA, ILL, IND, KSC, KY, LAF, MISS, MO, NDA, NEB, NY, OKLA, OS, PH, SDU, SMU, TENN, TEX, UNM, US, VDB, WIS, WVA). Over 1,200 annotated specimens were used and are available in these herbaria. For brevity, lists of exsiccatae are not given in this paper. After the range of distribution for the taxa was determined, a map of the area was divided into quadrats of approximately 150 × 150 miles each. Herbarium material was then assembled according to quadrat and measured or scored for length of leaf blade, width of leaf blade, leaf blade length × width, leaf blade length/width index, involucre length, shape score of the involucral bract tip, and leaf margin.

The leaf used for measurement was selected from the middle of the stem, and the head was selected from the center of the inflorescence from both herbarium specimens and population samples. The mean, standard deviation, and ranges were determined for each of the characteristics, both from population samples and from herbarium specimens.

DATA: figures 6.3.1 through 6.3.7; table 6.3.1.

REFERENCE: Jones, S. B. 1972. *Brittonia* 24: 28–45.

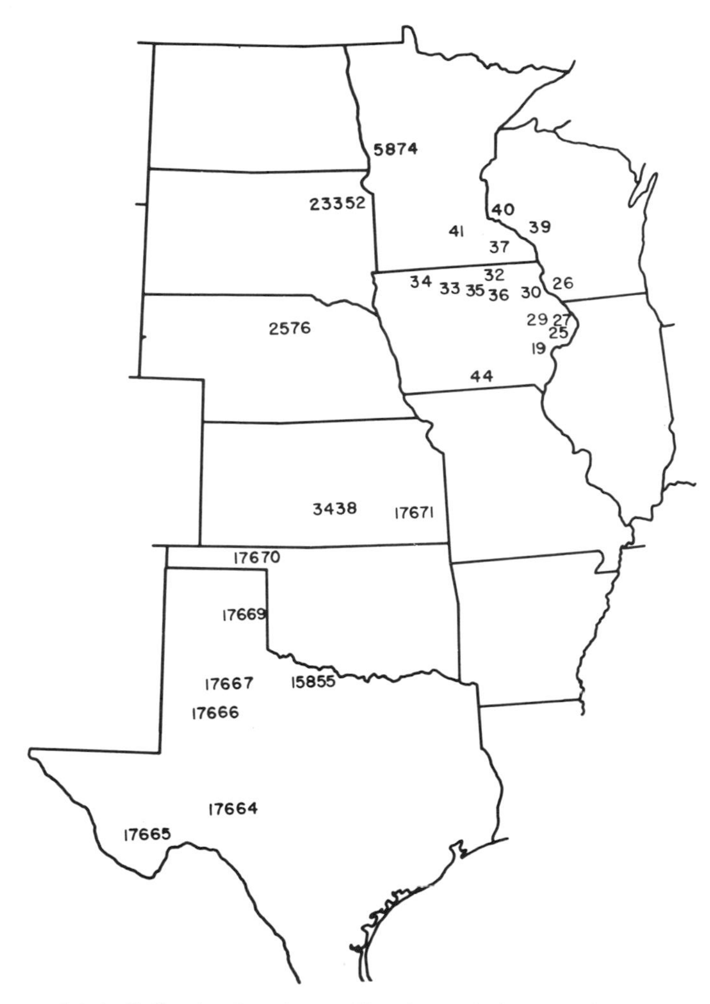

FIGURE 6.3.1 Collection locations of local population samples.

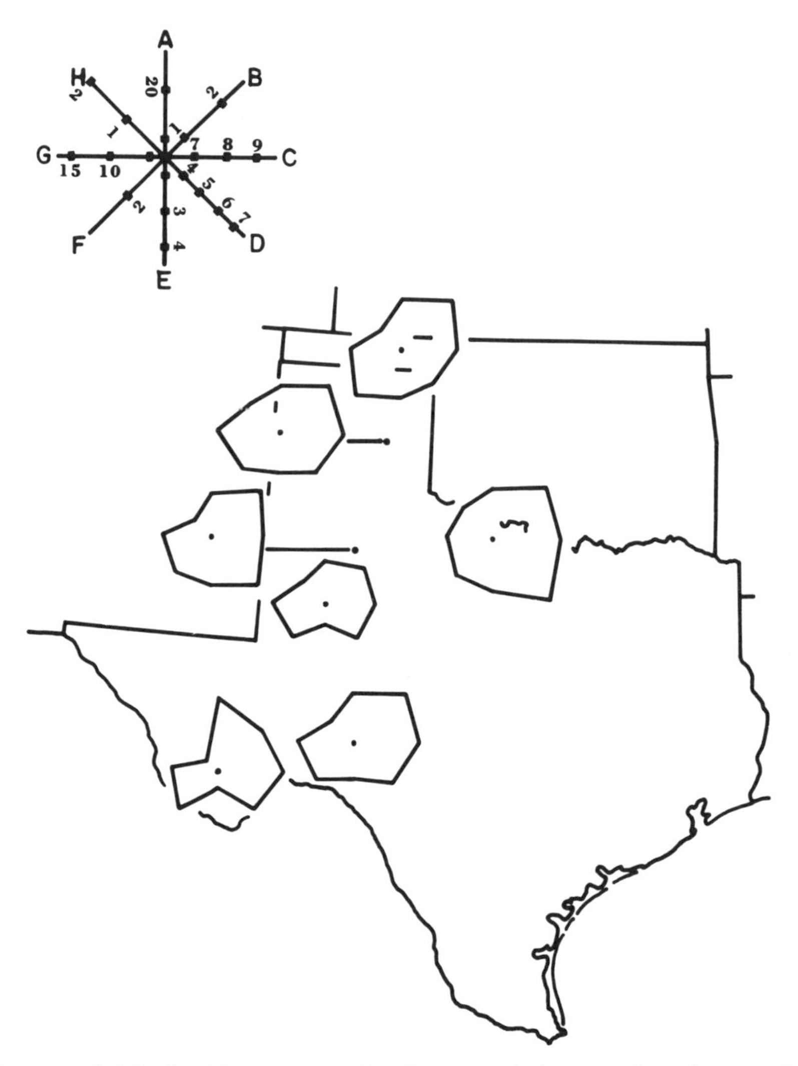

FIGURE 6.3.2 Graphic representation from population samples of means for: flowers per head (A); widest bract, mm (B); involucre length, mm (C); involucre width, mm (D); leaf margin score (E); bract tip shape score (F); leaf blade length, cm (G); and leaf blade width, cm (H). Means are presented in table 6.3.1.

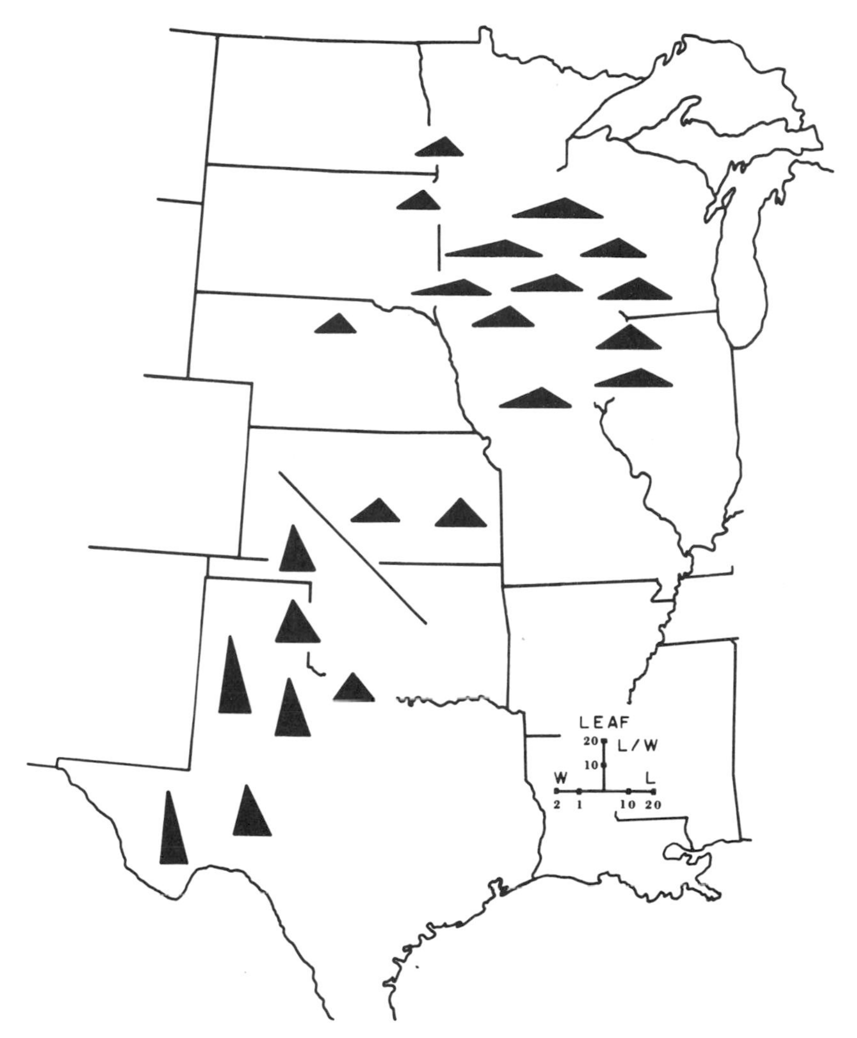

FIGURE 6.3.3 Graphic representation from population samples of means of leaf blade length and width measurements. The measurements are in cm. Means are presented in table 6.3.1.

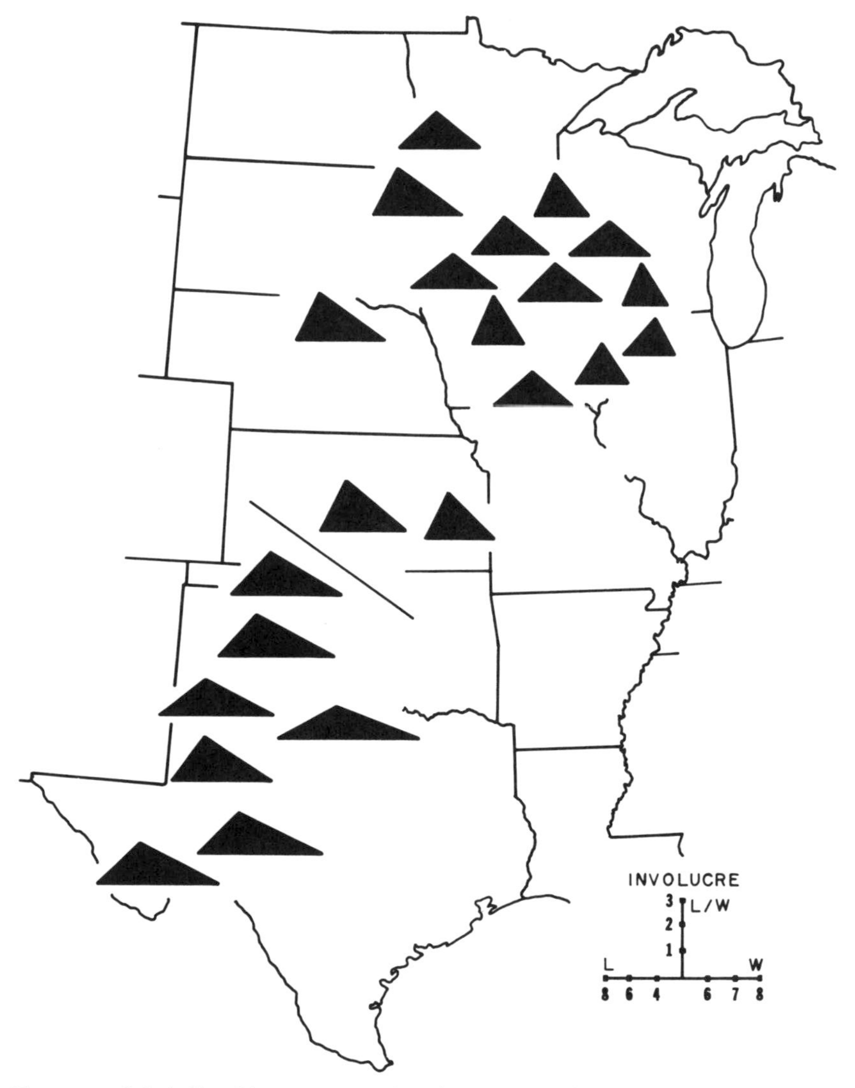

FIGURE 6.3.4 Graphic representation from population samples of means of involucre length and width measurements. The measurements are in mm. The means are presented in table 6.3.1.

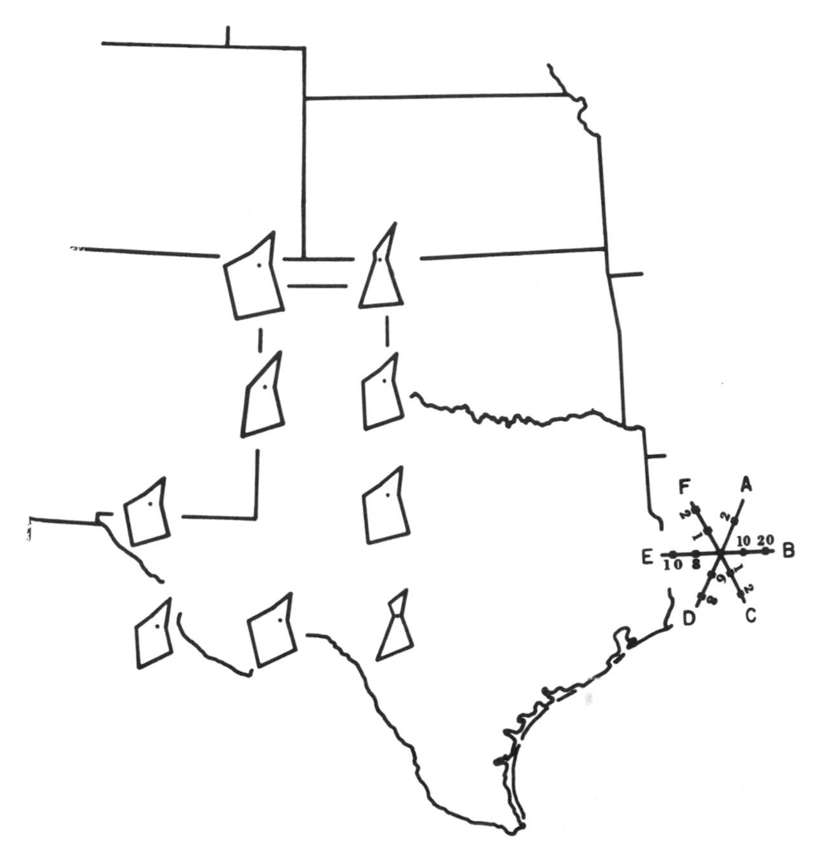

FIGURE 6.3.5 Graphic representation from herbarium specimens of quadrat means for: leaf margin score (A); leaf blade length × width (B); bract tip score (C); involucre length, mm (D); leaf blade length, cm (E); and for leaf blade width, cm (F). Means are presented in table 6.3.1.

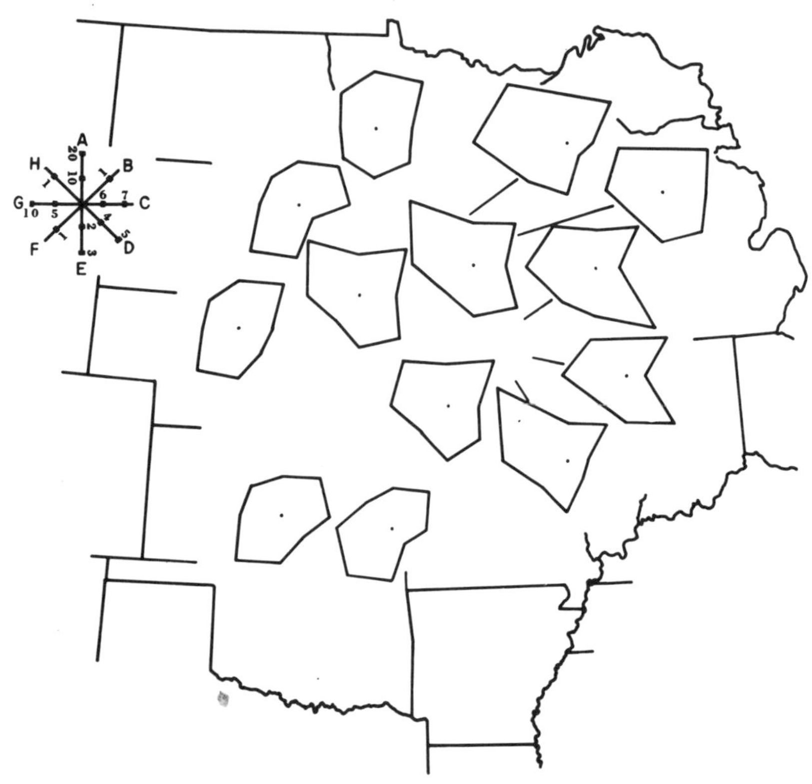

FIGURE 6.3.6 Graphic representation from population samples of means for: flowers per head (A); width of the widest bract, mm (B); involucre length, mm (C); involucre width, mm (D); leaf margin score (E); bract tip shape score (F); leaf blade length, cm (G); and leaf blade width, cm (H). Means are presented in table 6.3.1.

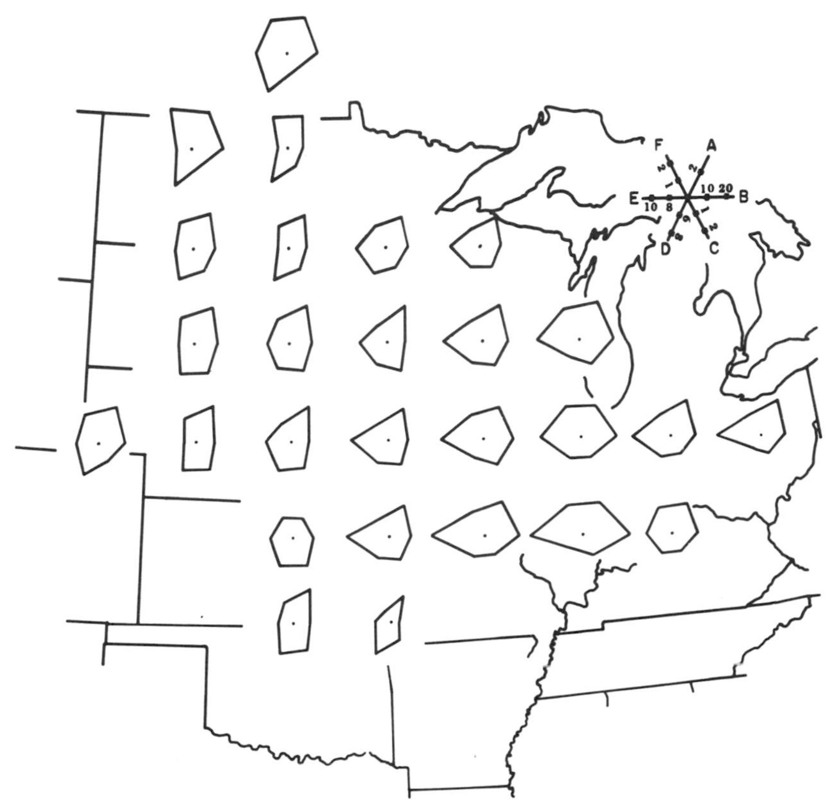

FIGURE 6.3.7 Graphic representation from herbarium specimens of quadrat means for: leaf margin score (A); leaf blade length × width (B); bract tip score (C); involucre length, mm (D); leaf blade length, cm (E); and for leaf blade width, cm (F). Means are presented in table 6.3.1.

TABLE 6.3.1.

Means and Standard Deviations of Ten Characteristics from Local Population Samples, N = 25

	Leaf Blade				Involucre			Phyllary		
Populations[a]	Length (cm)	Width (cm)	Length/Width Ratio	Margin Score	Length (mm)	Width (mm)	Length/Width Ratio	Width of Widest (mm)	Shape of Tips	Flowers per Head
17670	8.6 (1.6)	0.5 (0.2)	17.5 (4.4)	2.7 (0.6)	7.8 (1.1)	5.0 (0.6)	1.6 (0.3)	1.9 (0.3)	1.7 (0.5)	19.5 (1.8)
17669	10.1 (2.1)	0.7 (0.3)	15.8 (5.7)	2.5 (0.8)	8.1 (1.1)	5.0 (1.0)	1.6 (0.3)	1.8 (0.3)	1.8 (0.3)	18.8 (2.2)
15855	8.4 (2.3)	0.8 (0.3)	10.5 (2.8)	1.9 (0.3)	8.3 (1.1)	6.5 (0.5)	1.3 (0.3)	2.0 (0.3)	2.0 (0.2)	20.3 (2.1)
17667	8.6 (1.5)	0.4 (0.5)	30.8 (8.2)	2.7 (0.3)	7.7 (1.0)	5.6 (0.6)	1.4 (0.3)	1.9 (0.4)	1.8 (0.3)	17.5 (3.0)
17666	8.3 (2.1)	0.5 (0.7)	22.0 (5.8)	2.1 (0.8)	7.6 (0.6)	4.7 (0.6)	1.6 (0.3)	1.5 (0.3)	1.9 (0.3)	17.0 (2.3)
17664	9.7 (1.5)	0.5 (0.2)	19.5 (4.8)	2.5 (0.6)	8.3 (1.2)	5.2 (0.5)	1.6 (0.3)	1.8 (0.2)	2.0 (0.2)	18.5 (4.1)
17665	7.6 (1.2)	0.3 (0.1)	28.1 (7.8)	2.0 (0.8)	8.2 (0.7)	5.1 (0.7)	1.6 (0.3)	1.6 (0.2)	2.0 (0.0)	21.8 (1.9)
23352	7.0 (0.7)	1.0 (0.3)	7.2 (1.9)	3.0 (0.0)	7.5 (0.6)	4.0 (1.1)	1.8 (0.3)	1.5 (0.4)	1.8 (0.4)	17.2 (3.1)
2576	5.9 (1.4)	1.1 (0.3)	5.6 (1.3)	2.9 (0.2)	6.6 (0.9)	4.2 (0.4)	1.6 (0.2)	1.7 (0.2)	1.6 (0.5)	18.3 (2.2)
5874	7.7 (0.9)	1.2 (0.5)	6.9 (1.9)	3.0 (0.0)	6.7 (0.6)	4.9 (0.7)	1.4 (0.3)	1.8 (0.3)	1.3 (0.5)	21.3 (2.2)
3438	8.1 (1.2)	1.1 (0.5)	8.4 (2.7)	2.8 (0.4)	7.4 (0.5)	4.0 (0.5)	1.9 (0.3)	1.6 (0.3)	1.8 (0.4)	15.6 (2.4)
17671	10.2 (1.3)	1.0 (0.3)	10.9 (3.2)	3.0 (0.0)	6.8 (0.6)	3.9 (0.4)	1.8 (0.2)	1.5 (0.2)	1.8 (0.4)	17.3 (3.3)
34	10.9 (1.4)	2.1 (0.6)	5.5 (1.4)	3.0 (0.0)	6.7 (0.5)	5.2 (0.4)	1.3 (0.2)	1.8 (0.3)	1.0 (0.0)	18.2 (2.2)
33	10.0 (1.1)	1.5 (0.4)	7.3 (1.9)	3.0 (0.0)	6.3 (0.5)	3.6 (0.6)	1.8 (0.6)	1.7 (0.2)	1.5 (0.5)	16.6 (2.4)
41	15.3 (1.9)	2.6 (0.7)	6.1 (1.2)	3.0 (0.0)	6.7 (0.5)	4.4 (1.5)	1.4 (0.2)	1.7 (0.4)	1.3 (0.5)	17.7 (2.2)
25	12.2 (1.3)	1.4 (0.3)	9.2 (2.4)	2.8 (0.4)	5.7 (0.5)	4.5 (1.4)	1.4 (0.2)	1.5 (0.4)	1.0 (0.0)	14.1 (2.9)
32	10.7 (2.6)	2.0 (0.5)	5.8 (1.1)	3.0 (0.0)	6.8 (0.6)	4.8 (0.7)	1.4 (0.2)	1.9 (0.3)	1.3 (0.5)	15.4 (2.5)
36	10.3 (1.3)	1.3 (0.4)	8.5 (2.4)	2.9 (0.2)	5.8 (0.6)	3.5 (1.0)	1.7 (0.2)	1.6 (0.2)	1.6 (0.5)	19.5 (3.9)
35	9.9 (1.1)	1.2 (0.5)	9.6 (3.2)	3.0 (0.0)	6.4 (0.3)	3.9 (0.2)	1.7 (0.1)	1.5 (0.3)	1.5 (0.5)	17.6 (2.7)
27	14.5 (1.6)	2.0 (0.4)	7.4 (1.8)	3.0 (0.0)	5.7 (0.5)	4.1 (0.4)	1.4 (0.2)	1.5 (0.2)	1.0 (0.0)	13.4 (1.9)
40	15.8 (1.3)	2.3 (0.4)	7.2 (1.5)	3.0 (0.0)	6.3 (0.4)	3.7 (0.4)	1.7 (0.2)	1.6 (0.3)	1.4 (0.5)	17.9 (3.6)
29	12.2 (1.3)	1.6 (0.4)	7.9 (1.6)	3.0 (0.0)	5.5 (0.6)	3.5 (0.5)	1.5 (0.4)	1.5 (0.2)	1.2 (0.4)	11.9 (2.7)
39	11.3 (2.1)	1.7 (0.4)	7.1 (1.8)	3.0 (0.0)	6.7 (0.6)	5.1 (0.7)	1.3 (0.2)	1.7 (0.4)	1.0 (0.0)	17.4 (2.8)
44	11.9 (1.6)	1.6 (0.4)	7.4 (1.5)	3.0 (0.0)	6.6 (0.7)	5.0 (1.5)	1.3 (0.2)	1.8 (0.5)	1.0 (0.0)	16.2 (2.4)
19	12.4 (1.3)	1.9 (0.4)	6.9 (1.4)	3.0 (0.0)	6.0 (0.5)	4.0 (0.9)	1.5 (0.2)	1.5 (0.1)	1.0 (0.0)	15.4 (1.9)
26	13.4 (1.5)	1.6 (0.5)	8.7 (2.0)	2.9 (0.3)	6.0 (0.5)	3.6 (0.4)	1.6 (0.4)	1.6 (0.4)	1.4 (0.5)	14.7 (2.4)
37	12.8 (1.2)	2.5 (0.7)	5.4 (1.2)	3.0 (0.0)	6.5 (0.4)	5.2 (0.6)	1.3 (0.1)	1.7 (0.3)	1.0 (0.0)	16.6 (2.6)
30	13.1 (1.7)	2.1 (0.7)	6.5 (1.2)	3.0 (0.0)	5.8 (0.5)	4.0 (0.6)	1.5 (0.2)	1.5 (0.1)	1.0 (0.0)	14.0 (1.7)

[a] See figure 6.3.1.

Case Study Exercise 4 (Liliaceae)

PROBLEM: Provide a classification for the five taxa (A–E), using any or all of the following categories: species, subspecies, variety, form.

MATERIALS AND METHODS: The study was conducted with fresh material collected in the field. This approach is particularly necessary with the taxa under consideration because the critical features are obscured by pressing and drying. At the conclusion of the study, herbarium specimens were examined to fill in the gaps of distribution for each taxon and to gain a more complete picture of the patterns of variation. (Thanks are expressed to the curators of the herbaria from which these specimens were borrowed: DUKE, FSU, GA, NCSU, NCU, PENN, PH, TENN, US, VPI, WVA.)

The collection sites were selected on the basis of observations of herbarium specimens, as well as consideration of areas critical to the general distribution of the various variants. Many of these sites had to be visited twice—first for the flowers, and then later for the mature capsules. (Thanks are expressed for the assistance given by Dr. Roland M. Harper in pointing out sites for collecting in Alabama; by Mr. R. L. James for directions to a number of very interesting types near Johnson City, Tennessee; and by the late Mr. Robert Sparks Walker who located a large population near Chattanooga, Tennessee.)

Specimens were kept fresh by refrigeration until they could be evaluated. The leaves were pressed and later measured and a ratio was calculated by dividing width into length. The length was measured from the apex to the point of basal widening of the leaf "blade." The length of the petiole-like base was measured from the point of sheathing of the outer leaf on the stem to the point of basal widening of the "blade." The leaf "blades" were divided into sixteen equal segments by length, and the segment where the widest point fell, counting from the base, was recorded for each. A ratio of the largest leaf to the length of the peduncle was also calculated. Measurements of sepal, petal, ovary, style, and stigma were taken, and the characteristics of these organs were recorded. Specimens past flowering were considered separately, and particular attention was paid to the orientation of the fruiting peduncle.

The habitat was noted with respect to the associated flora, proximity to streams, and other features. The presence or absence of runners or stolons was observed, as was the relative number of flowering specimens in a population. Unusual sites were considered in more detail.

Measurements were kept separately for each plant, and population means were calculated later for comparison.

The first chromosome counts were made from squash preparations of root tips from plants grown in pots under greenhouse culture, but later it was found that a similar preparation from the tip of the actively growing stolon was far superior to root-tip preparations, as the stolon is much larger, has many more divisions, and has soft tissue that is easily prepared for examination. Although an active stolon can be collected from field specimens during flowering and even until just before capsules mature, the earlier stages before bulblet formation commences are preferred for slide preparation.

It was found that a pretreatment with oxyquinoline for twelve to twenty-four hours (Saylor. 1961. *Silvae Genetica* 10: 77–84) shortened the chromosomes and thus greatly improved the preparations. Since this technique was not used for the field-collected material (stolons) from which most of the counts were made, it is merely pointed out that the pretreatment should be performed in further studies of the chromosomes. This procedure would be particularly helpful for analysis of karyotypes.

The procedure that was used to prepare slides was suggested by Saylor (1961). The material was fixed in the field in a 3:1 (ethanol: acetic acid) fixative. Hydrolysis was carried out with a solution of 1 part stock HCl to 9 parts aceto-orcein; staining was subsequently carried out with neutral aceto-orcein.

DATA: figures 6.4.1 and 6.4.2; table 6.4.1.

REFERENCE: Parks, C. R. and J. W. Hardin. 1963. *Brittonia* 15: 245–259.

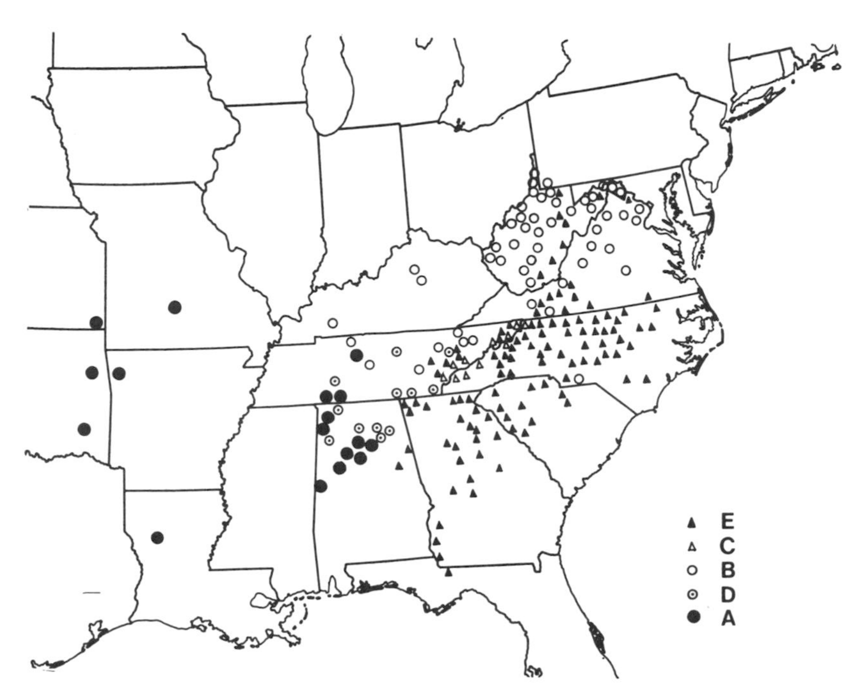

FIGURE 6.4.1 Distribution of the five taxa A–E.

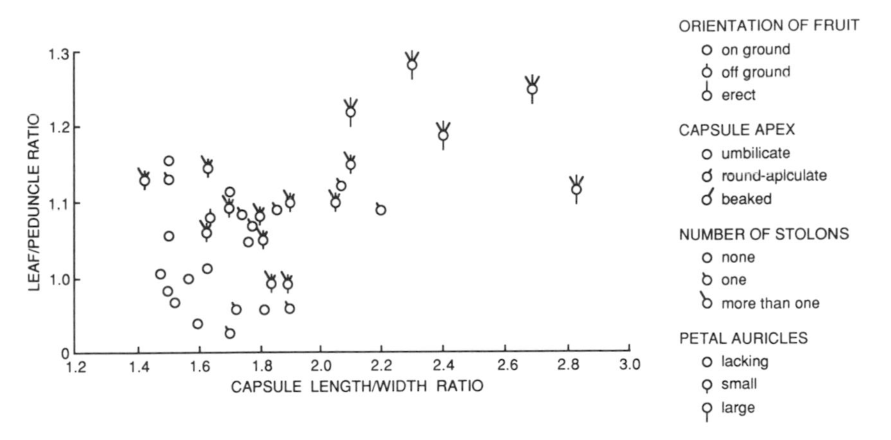

FIGURE 6.4.2 Pictorialized scatter diagram showing relationships of taxa A–E by the correlation of diagnostic features. The metroglyphs at the upper right, with four long rays, represent Taxon A; those at the lower left, without rays, represent Taxon E; a single short ray at the top of a metroglyph distinguishes Taxon C; metroglyphs intermediate in position and with one long and three short rays represent taxa B and D.

TABLE 6.4.1.

Summary of Morphological and Chromosomal Characteristics of Taxa A–E

Character	Taxon A	Taxon B	Taxon C	Taxon D	Taxon E
Chromosome number $2n$ =	24	48	24	48	24
Stolon number	>1	>1	1	>1	0
Petal anthocyanin	intense on veins abaxial side	variable: light to deep	variable: slight to nearly absent	considerable pigmentation[a]	variable: veins often intense
Sepal anthocyanin	intense on abaxial side	variable: light to deep	variable: usually reduced to absent	considerable pigmentation[a]	variable: usually intense on abaxial side
Flecking (dark spots on perianth)	absent	absent to slight	variable: few to many but always present	[b]	variable: usually slight to absent
Pale spot on adaxial side of perianth	absent	vague or small; absent in 90%	variable: small to large; always present	[b]	small to medium size; usually present
Green on abaxial side of perianth	absent	rare or slight if present	present	[b]	absent
Anther color	yellow	usually yellow, sometimes brown to lavender	rarely yellow; light to deep purple	usually yellow	rarely yellow; usually lavender to brown-purple
Pollen color	yellow	yellow; rarely yellow-brown	rarely yellow; usually brown or purple	[b]	rarely yellow; usually brown or purple
Style color at flowering	yellow-green	green-yellow	yellow to greenish yellow	green-yellow	yellow
Auricles	present and large	present	absent; margin irregular	present or absent[a]	absent

Style persistence	until fruit matures	often 1/2 withered by fruit maturity	withers soon; basal 1–2 mm may persist	often 1/2 withered by fruit maturity	withers soon after fruit matures
Form of stigma	swollen, lobes short and erect	swollen, lobes long and divergent	slender, lobes long and divergent	swollen, lobes short and divergent	slender, lobes stout and divergent
Capsule shape (apex; long section)	strongly beaked (rostrate)	round to truncate or apiculate	umbilicate to rarely truncate	distinctly apiculate	distinctly indented (umbilicate)
Capsule shape (cross section)	round to slightly 3-angled	3-angled to distinctly triangular	3-lobed; the lobes rounded	distinctly triangular	3-lobed; the lobes rounded
Orientation of capsule	held erect	variable, but usually well off ground	reclining on ground or rarely just above	variable, but well off ground	reclining on ground

[a] Based on few observations.
[b] Not observed.

Case Study Exercise 5 (Ericaceae)

PROBLEM: Provide a classification of taxa A and B, using any or all of the following categories: species, subspecies, variety, form.

MATERIALS AND METHODS: *Geographic range and ecology:* specimens from GH, NCSC, NCU, US, USCH, and USNA were examined early in the research. Over 1,000 voucher specimens for examination and documentation were collected throughout the geographical range of Taxa A and B during the field studies and were deposited at NCSC and USCH. Vegetation and site data were also recorded for ecological analysis. A site index listing location and approximate population size was prepared to document the geographical range and distribution of Taxa A and B (Kirkman. 1987a. Ph.D. Diss. Raleigh: North Carolina State University; Kirkman. 1987b. *Hort. Crops Res. Ser.* 74. Raleigh: North Carolina State University). Many of these sites as well as a few additional sites were also documented by Ballington et al. (1980, 1983, 1986. *Hort. Crops Res. Ser.* 51, 60, 71).

Kirkman et al. (1989. *Bull. Torrey Bot. Club* 116: 114–133) have presented a description and classification of the habitats and a ranking of plant associates for Taxa A and B. In addition, DECORANA ordination (Hill. 1979a. Ithaca: Cornell Univ. Sect. Ecol. Syst.), TWINSPAN numerical classification (Hill. 1979b. Ithaca: Cornell Univ. Sect. Ecol. Syst.), and step-wise linear regression (SAS Institute. 1982. Cary, North Carolina) were used to identify many of the important environmental relationships of the taxa, especially fire. A comprehensive and systematic study of the ecology and phytosociology of Taxa A and B, recognition of habitat differences/similarities between them, and the role of the environment in morphological variation, were essential elements in the determination of the rank of the two taxa.

The distributions of taxa are shown in figure 6.5.1. Taxon A is primarily in open pinelands of eastern North and South Carolina, with the heaviest concentration in southeastern North Carolina and northeastern South Carolina. It occurs in the Outer Coastal Plain of South Carolina and sporadically inland to the fall zone just below the North Carolina line. In North Carolina it occurs throughout most of the Coastal Plain, Sandhills, and occasionally into the eastern Piedmont. Taxon B is endemic to Lexington County in the Sandhills of South Carolina, but the record of a now-extinct population of Taxon A in nearby Kershaw County reduced the geographical separation of the two species from approximately 130 km to less than 60 km (Kirkman 1987a, 1987b).

Sandhills populations of Taxa A and B occur in disjunct Coastal Plain communities along streams, as in Cheraw State Park, South Carolina, or in upland sites with perched water tables, moisture-holding soils (increased clay content), or seepage zones, such as those described by Wells and Shunk (1931. *Ecol. Monogr.* 1: 465–520). Analysis of the population at Shealy's Pond, the type locality for Taxon B, and the other Lexington County populations was especially important in evaluating the ecological and taxonomic status of the taxa. In the ordinations (Kirkman et al. 1989), Shealy's Pond was an extreme endpoint but not a statistical outlier; highly organic pocosin stands from the Outer Coastal Plain were its closest associates. Despite its location in the Sandhills and its unique vegetational composition (1984. *Nature Conservancy News* 34:27), the physical characteristics and physiognomic composition of Shealy's Pond are quite similar to those of pocosin areas in southeastern North Carolina. The other two Lexington County stands used in ordination and numerical classification were not extremes and were closely associated with other Sandhills stands of Taxa A and B. While most populations of Taxa A and B are subjected to the stress of periodic seasonal drought, the hillside seepage bog at Shealy's Pond does not become dry even in extreme drought years such as 1986.

Field collections, common garden plots, and transplant experiments: The geographical and ecological ranges of Taxa A and B were surveyed using existing herbarium collections, reports of other field botanists, and the authors' own collections. Preliminary field work showed that individual clones of Taxa A and B showed a high degree of phenotypic plasticity in their response to the environment. In an attempt to minimize the effect of environment, as well as to study the adaptability of suspected ecotypes or geographical variants, common garden plots and transplant experiments were established in 1979, using plants grown from rooted leafy stem cuttings (Kirkman 1987a). Container plants grown from rooted cuttings supplemented these studies and were used in crossability studies.

Particular attention was given to differences and/or similarities in reproductive morphology of all material examined. Only mature leaves were sampled for statistical analysis of vegetative morphology; every third leaf was measured on all branches to insure representation. Because Taxa A and B are genuinely evergreen, this procedure also insured representation from more than one growing season.

Forty-three asexually propagated clones, chosen to represent the geographical, ecological, and morphological diversity of Taxa A and B, were used in the replicated, randomly ordered common garden plots at Castle Hayne,

North Carolina (Kirkman 1987a). The 43 clones included 13 selected from Lexington County, South Carolina, including one from the type locality of Taxon B, and the first F_1 hybrid (made by Ballington) of Taxa B and A. Plants were rated periodically for survival (initial failures were replaced in 1980), vigor, disease, and growth habit (Kirkman 1987a; Kirkman and Ballington. 1985. *Hort. Science* 20: 1138–1140).

Common garden plots were used for phenological studies, floral and fruit counts and measurements, and comparative studies of vegetative morphology. In 1982 voucher specimens suitable for statistical analysis of taxonomic measurements were collected and preserved from the common garden plots. Selected plants were dug and root systems examined for comparison with field studies.

Transplant experiments were established at five locations in 1979 and were monitored through 1983 (Kirkman 1987a). The five sites included three diverse Taxon A populations and two Taxon B populations (the type locality at Shealy's Pond and an upland location). Clones used in the experiments were chosen for geographical and morphological representation. Rooted cuttings of five clones, randomly ordered, were planted in four repetitions and spaced 1 m apart at each site. The plantings were monitored periodically for survival, vigor, disease, morphological changes, and flower and fruit set.

Transplants at the five sites varied in survival rates according to specific microclimate in the randomization and clonal response (Kirkman 1987a). Surviving clones at Shealy's Pond exhibited a significant increase in leaf size and showed heavy disease pressure in the moist shaded environment. Transplants at the upland site in Lexington County, South Carolina, and at the site in Montgomery County, North Carolina, showed excellent survival for all clones—even after five years. These transplanted populations showed a decrease in leaf size of the larger-leaf clones that had been brought in from lowland Lexington County sites, but there was essentially no change in typical Taxon A clones or upland Lexington County clones (Taxon B).

Results of these studies confirm that there is a genetic basis for much of the phenotypic variation in Taxa A and B, but that variation is limited to vegetative characters, specifically leaf shape and size (Kirkman 1987a). Analysis of data from the common garden plots was important in determining the effect of environment on the overlap and separation of leaf characters. Data from transplant experiments were important in assessing possible habitat separation; final determination was based on phytosociological studies and habitat analysis, as discussed in Kirkman et al. (1989).

DATA: figures 6.5.1 through 6.5.5; tables 6.5.1 and 6.5.2.

REFERENCE: Kirkman, W. B. and J. R. Ballington. 1990. *Syst. Bot.* 15: 679–699.

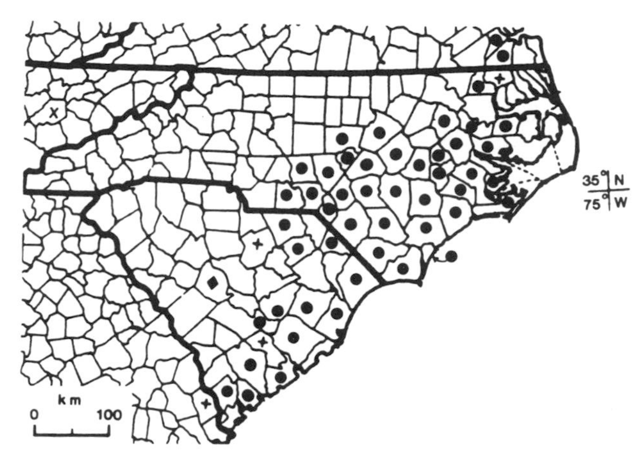

FIGURE 6.5.1 Distributions of taxa A and B. Blackened circles, collections of Taxon A made during the current study; pluses, collections of Taxon A made previously, but not refound in this study; X, unfounded report by Blauch (1970. Ph.D. dissertation, West Virginia University, Morgantown), probably neither Taxon A nor B; diamond, collections of Taxon B endemic to Lexington Co., South Carolina.

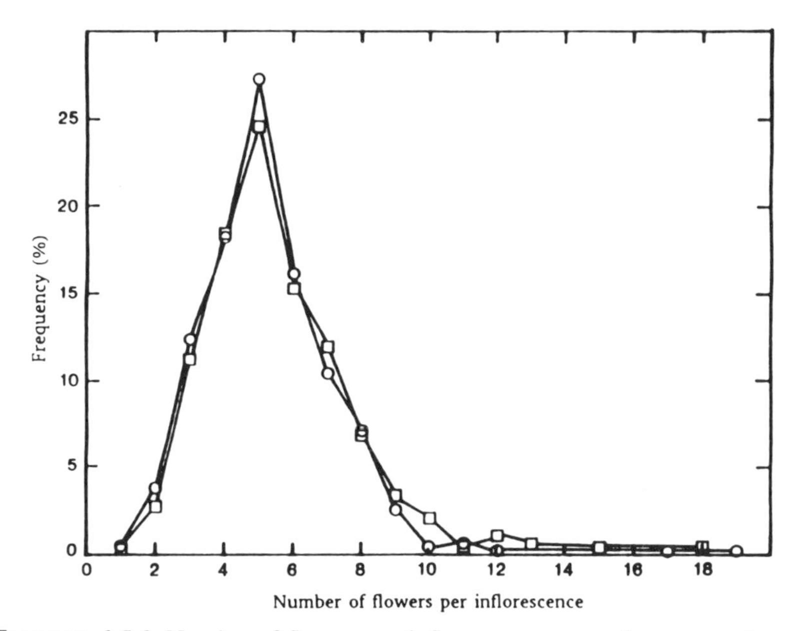

FIGURE 6.5.2 Number of flowers per inflorescence versus frequency of occurrence in common garden plots at Castle Hayne, North Carolina, and in container culture/research plots of taxa A and B at Raleigh, North Carolina. Circles = Taxon A ($\bar{x}$ = 5.23 ± 1.88; N = 759). Squares = Taxon B ($\bar{x}$ = 5.55 ± 2.21; N = 468).

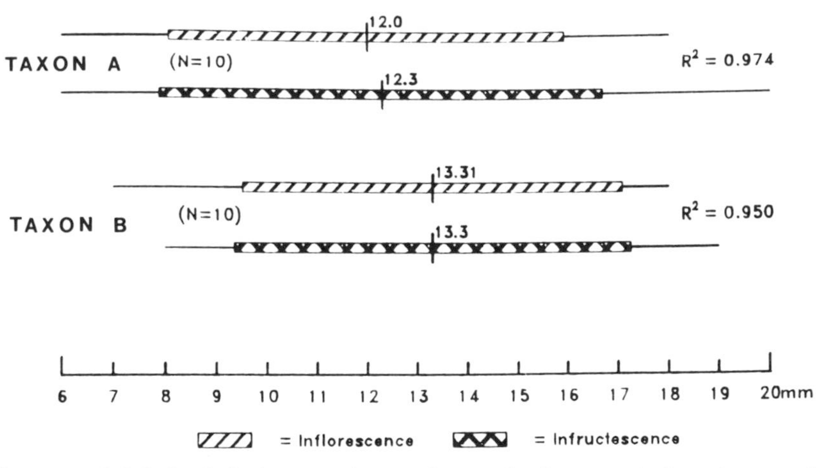

FIGURE 6.5.3 Statistical comparisons of reproductive morphology in taxa A and B: paired comparisons of inflorescence/infructescence lengths in Taxon A (K-81-25) and B (NC 78-13-1).

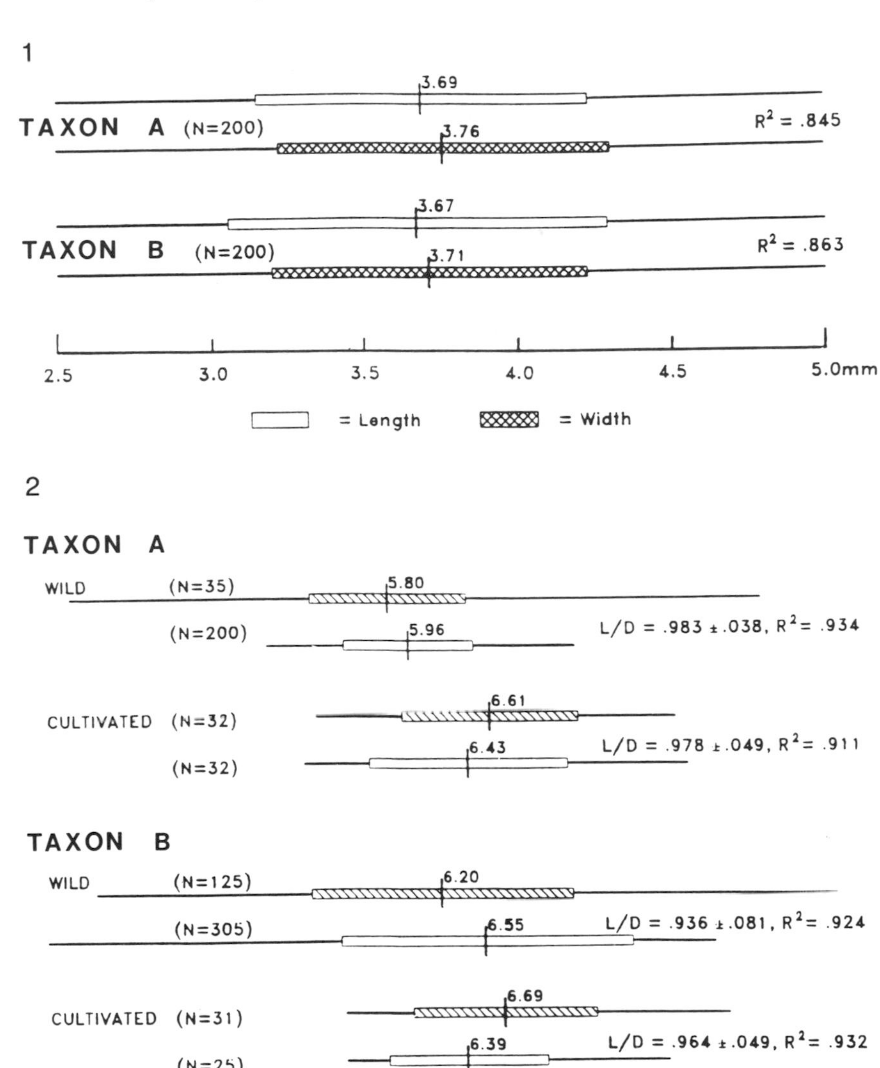

FIGURE 6.5.4 Statistical comparisons of reproductive morphology in taxa A and B. 1: corolla measurements for taxa A and B. 2: fruit measurements for wild and cultivated samples of taxa A and B.

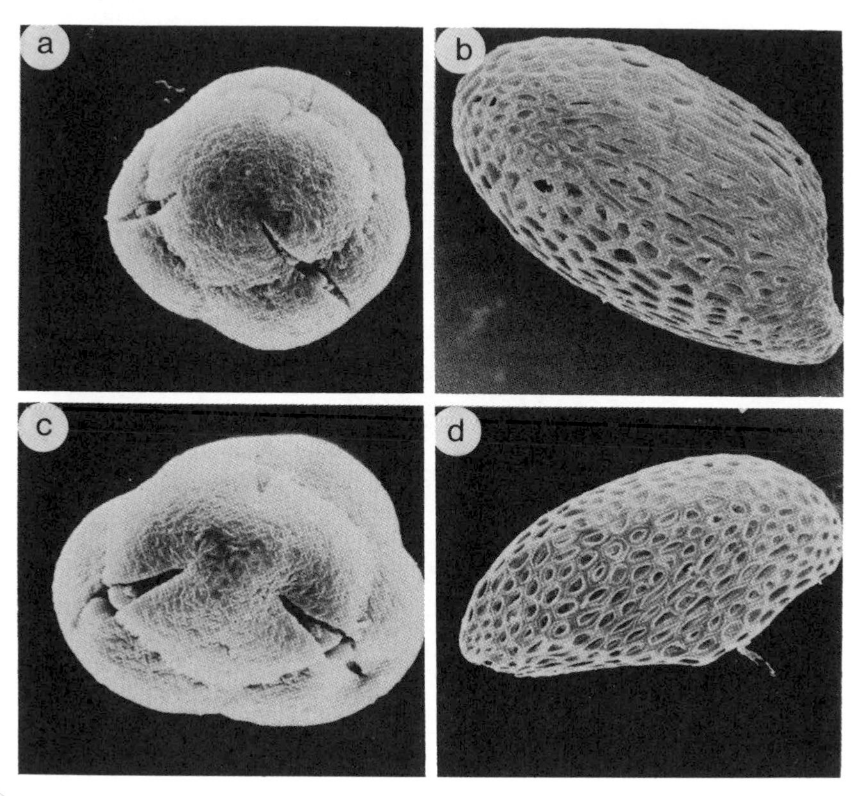

FIGURE 6.5.5 Pollen tetrads (1 and 3) and seeds (2 and 4) in taxa A and B. Taxon A, 1 and 2 (K-79-35, K-79-41). Taxon B, 3 and 4 (B-51-10, K-79-4). Magnification: 1 and 3 = $\times 1900$; 2 = $\times 80$; 4 = $\times 70$.

TABLE 6.5.1.
Comparison of Foliar Morphology of Taxa A and B

Character	Taxon A N = 2827	Taxon B N = 1319
Blade length (mm)	(2)3–18(25)[a]	(4)7–35(63)[a]
Blade width (mm)	(1)2–9(15)	(2)3–20(36)
Petiole length (mm)	(0.5)1.0–2.0(2.5)	(0.5)1.0–3.0(4.5)
Ratio blade length to blade width	(1.0)1.5–2.7(4.5)	(0.9)1.4–2.6(4.0)
Ratio blade length to petiole length	(2)4–12(20)	(3)5–15(40)
Leaf shape	6.3% ovate 13.5% ell-ovate 60.8% elliptic 12.1% ell-obovate 6.3% obovate 1.0% others	0.0% ovate 7.7% ell-ovate 35.5% elliptic 28.4% ell-obovate 31.2% obovate 1.1% others
Leaf apex	0.0% acuminate 54.3% acute 39.5% obtuse 6.3% rounded	0.7% acuminate 41.2% acute 49.5% obtuse 8.6% rounded
Leaf base	0.6% narrow-cuneate 54.6% acute 37.0% obtuse 7.7% rounded	12.3% narrow-cuneate 65.4% acute 20.9% obtuse 1.4% rounded

[a]Figures in parentheses represent total range; inner figures represent modal range of values (≥ 90% of data).

TABLE 6.5.2.
Comparison of Leaf Shapes (%) of Taxon A and B Seedlings

Taxon A N = 50		Taxon B N = 50	
Ovate	16	Ovate	10
Elliptic-ovate	22	Elliptic-ovate	14
Elliptic	36	Elliptic	38
Elliptic-obovate	20	Elliptic-obovate	18
Obovate	6	Obovate	20

NOTE: Seedlings recorded at age one year.

Case Study Exercise 6 (Lemnaceae)

PROBLEM: Provide a classification of 21 species (Taxa A–U) into genera.

MATERIALS AND METHODS: One hundred eighty-six clones of 21 taxa (A–U) were obtained from natural sources and from established cultures in other laboratories. All clones were cultured axenically on defined media. A list of the experimental clones appears in table 6.6.1 (for complete collection data, see McClure. 1964. Ph.D. Diss. Austin: University of Texas). When different clones of the same taxon were obtained, a "key" clone was selected for each taxon—i.e., one which for the purpose of these investigations would serve as a representative of the taxon and with which the other clones would be compared. These key clones were selected on the basis of their morphological similarity to the original descriptions of the species and in most instances were compared with the holotype. In the absence of holotypes, specimens annotated by previous monographers (Hegelmaier. 1868. *Die Lemnacean: Eine Monographische Untersuchung*. Leipzig: Wilhelm Engelmann; Daubs 1965. *Illinois Biol. Monogr.* 34. Urbana: University of Illinois Press) were consulted. The flavonoid chemistry of these key clones was compared with that of all other clones available in culture.

The clones available for this investigation represent 21 of the 29 taxa recognized in the most recent monographic treatment of the family (Daubs 1965). Of the eight taxa not available in culture, five are known from single collections made over a century ago. The remaining five taxa are rarely collected and limited in distribution; living materials could not, therefore, be obtained despite numerous inquiries. Aseptic fronds were obtained by treatment with 10% Clorox and transferred to tubes of autoclaved 1/3 × Hutner's medium (Hutner. 1953. In Loomis, ed., *Growth and Differentiation in Plants*. Ames: Iowa State College Press) containing 1% sucrose (1 g per 100 ml), henceforth referred to as medium HS.

All stock cultures and experimental plants were grown in a controlled environment room with a constant temperature of 21° C and an 18-hr photoperiod. Illumination was supplied by horizontal banks of alternate warm-white and cool-white fluorescent lamps. Stock cultures were maintained on HS medium solidified with agar (1.25 g per 100 ml) at approximately 200 ft-c. Cultures for chemical or morphological analysis were grown routinely at a light intensity of approximately 1,000 ft-c with a temperature inside the flask of 26°C during the light period, dropping to 21°C during the dark period.

Because some clones exhibited a growth rate differing from other clones of even the same taxon, and because wide variations in growth characteristics exist among the taxa (Landolt. 1957. *Ber. Schweiz. Bot. Ges.* 67: 271–410), the plants were collected when they covered the surface of the medium, or with the slowest growing plants, after 28 days of growth.

Plant material adequate for single two-dimensional chromatograms (approximately 1 g of fresh material) could be obtained in most instances from cultures produced in a single 125-ml Erlenmeyer flask containing 50 ml of HS medium. Mass cultures of the key clones were grown for isolation and further analysis of individual flavonoids in 2,500 ml low-form culture flasks (Pyrex model 4422), containing 1 liter of HS medium.

Collections for chromatographic investigations were harvested by pouring the contents of the culture into a fine-mesh wire basket, washing the fronds thoroughly with distilled water, then placing them on sheets of absorbent paper. The collections were dried in a 50°C forced-air oven for 2 hr.

To determine the degree of reliability of the chromatographic patterns of the flavonoids, clone 7003 of Taxon K and clone SOH of Taxon B were subcultured under 62 different regimes (McClure and Alston. 1964. *Nature* 201: 311–313) and the resulting cultures were examined chromatographically.

Two-dimensional chromatograms were prepared from methanolic extracts of each clone grown under controlled conditions. Each clone was examined on at least three different occasions. About 0.1 g of dried plant material was placed in a small shell vial and extracted in 5 ml of 0.5% HCl in methanol for 12 hr at room temperature. The extracts were filtered, dried without heat, and taken up in a small amount of methanol and applied to a sheet (46 × 57 cm) of Whatman No. 3MM chromatographic paper. The chromatograms were developed by descending chromatography; first in tertiary butyl alcohol:acetic acid:water (3:1:1 v/v; henceforth referred to as 3:1:1) and then in acetic acid:water (15:85 v/v; referred to as 15% HOAc).

From 1.5 to 10.0 g of dried material from each taxon were used to prepare 20 or more two-dimensional chromatograms for isolation of individual flavonoids. The chromatograms were examined in ultraviolet light and marked; areas representing a single compound were cut out and combined. These compounds were eluted in either neutral methanol or 0.1% HCl in methanol in order to disclose the possible presence of anthocyanins. After 12 hr on a rotary shaker, the eluates were filtered, reduced in volume in vacuo, and each applied as a single streak across one-half the width of a full-size sheet of Whatman No. 3MM paper. These streaks were developed in either 3:1:1, 15% HOAc, or various other proportional mixtures of these solvents selected to facilitate purification of the flavonoid. Rechromatography was continued until only one band was resolved when checked in various solvents.

Absorption spectra of standard compounds and of eluates from paper chromatograms were measured in absolute methanol on a Beckman Spectrophotometer Model DB. Spectra of the crystalline flavonoid standards were obtained in many instances. In addition to a methanolic spectrum, spectra were obtained in sodium ethoxide, sodium acetate, aluminum chloride, and sodium acetate–boric acid solutions (Jurd. 1962. In Geissman, ed., *The Chemistry of Flavonoid Compounds*. New York: Macmillan).

A portion of the total methanolic eluate used for isolation and purification was hydrolyzed in 6 N methanolic HCl for 2 hr at 82°C to detect glycoflavones. These conditions hydrolyze O-glycosidic linkages, leaving intact the C-glycosidic linkages of glycoflavones. The hydrolytic products of the O-glycosyl flavonoids were obtained by heating the rechromatogrammed eluates for 30 min in 0.5 N methanolic HCl at 82°C. The resultant aglycones were chromatographically purified, and they were analyzed by absorption spectrometry and by co-chromatography with authentic samples.

As an example of the manner in which the compounds in figure 6.6.1 were identified, Compound 2 will be considered. On the two-dimensional chromatograms this spot had an R_F of ca. 0.55 in 3:1:1 and of ca. 0.35 in 15% HOAc. This result indicates that Compound 2 was not a common aglycone, since it would then have a higher R_F in 3:1:1 and a much lower R_F in 15% HOAc (Seikel. 1962. In Geissman, ed., *The Chemistry of Flavonoid Compounds*. New York: Macmillan). When sprayed with Benedict's reagent, the spot was yellow in ultraviolet light, indicating the absence of ortho-dihydroxyl groups. Hydrolysis in 6 N methanolic HCl for 2 hr yielded a mixture of the original compound and a second compound that has a higher R_F value in both solvents (Spot 23, figure 6.6.1). These are characteristics of certain glycoflavones.

When the rechromatographed and purified eluate was examined spectrally in methanol, the Band I maximum was at 333 mμ (figure 6.6.3). A single well-defined peak in Band II at 269 mμ further supported the absence of ortho-dihydroxyl groups in ring B. In sodium ethoxide the shift of 61 mμ of Band I indicated that a 4′ hydroxyl was present. With sodium acetate, both Bands I and II shifted, confirming the presence of a free 4′ hydroxyl and indicating that the 7-hydroxyl position was not glycosylated. The 5-hydroxyl position was shown to be free by the bathochromic shift of Bands I and II upon the addition of aluminum chloride. Therefore, the compound has free hydroxyls at positions 4′, 5 and 7, yet glycosidic groups are implicated because of the chromatographic behavior. These spectral and chromatographic characteristics are similar to those reported for vitexin (Jurd 1962; Seikel 1962). Authentic samples of vitexin yielded identical chromatographic and spectral data, and the compound was identified as vitexin. Vitexin also yields some saponaretin (compound 23) upon prolonged hydrolysis, as does compound 2.

Similar methods were applied in the structural determination of all flavones, glycoflavones, and flavonols detected within this family. The anthocyanins are not easily identified by these techniques, and they are not stable in the alkaline diagnostic reagents sodium ethoxide and sodium acetate (Jurd 1962). These compounds have been previously investigated, however, and the anthocyanin of Taxon B (compound 8, figure 6.6.1) is reportedly petunidin-3-5-diglucoside (Ng, Thimann, and Gordon. 1964. *Arch. Biochem. and Biophys.* 102: 92–95). The major anthocyanin of Taxon K (compound 11) is cyanidin-3-glucoside (Ng and Thimann. 1962. *Arch. Biochem. and Biophys.* 96: 336–339). Co-chromatography in multiple solvent systems confirmed the identity of compounds 8 and 11.

DATA: figures 6.6.1 through 6.6.3; tables 6.6.1 through 6.6.3.

REFERENCE: McClure, J. W. and R. E. Alston. 1966. *Amer. J. Bot.* 53: 849–860.

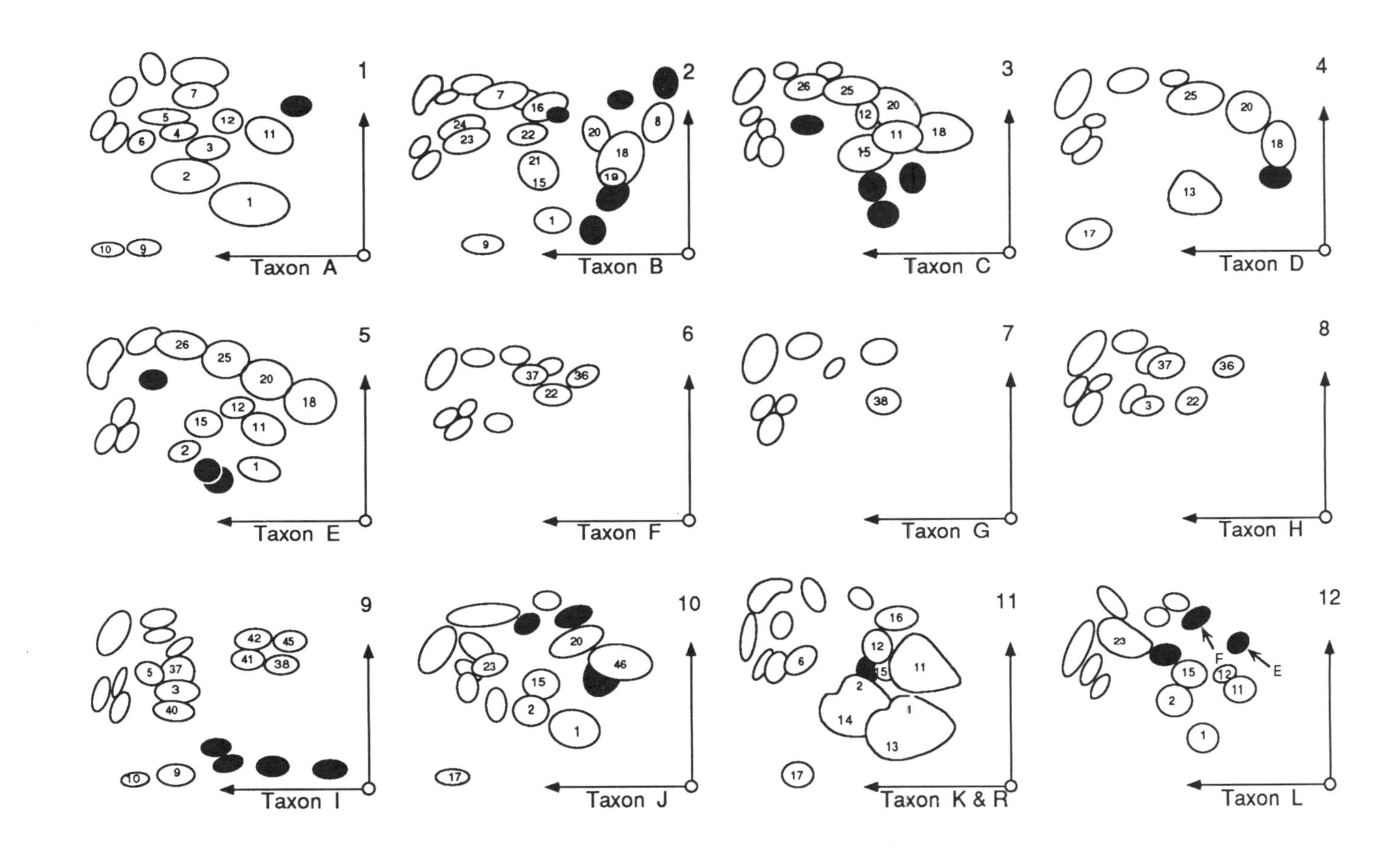
1
Taxon A
2
Taxon B
3
Taxon C
4
Taxon D
5
Taxon E
6
Taxon F
7
Taxon G
8
Taxon H
9
Taxon I
10
Taxon J
11
Taxon K & R
12
Taxon L
F
E

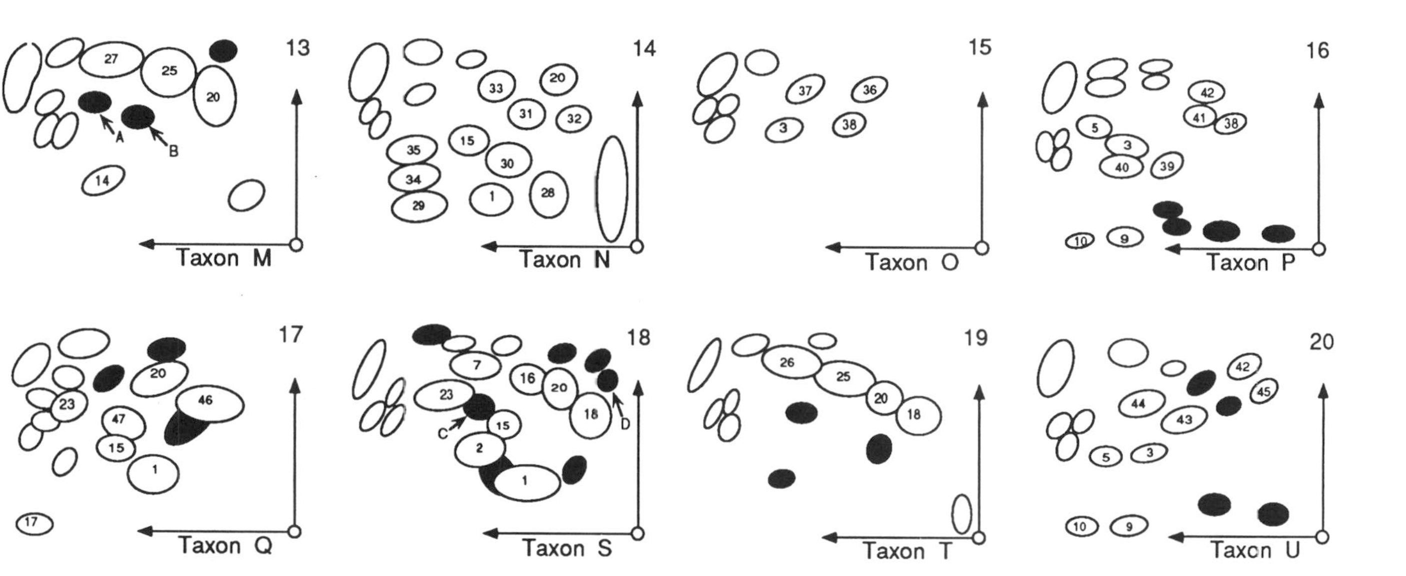

FIGURE 6.6.1 Two-dimensional chromatographic patterns of methanolic extracts of taxa A–U. Spots designated by numbers 1 through 47 are flavonoids, and they are characterized in table 6.6.2. Darkened spots are considered to be flavonoids on the basis of their chromatographic characteristics, but these have not been further identified. Undesignated spots represent unidentified substances which appear light blue in ultraviolet light; they sometimes give a modified color reaction with ammonia or Benedict's reagent and are not considered to be flavonoids.

A

Aglycones	Substituents R	R′	R″	R‴
apigenin	H	H	H	H
luteolin	OH	H	H	H
kaempferol	H	OH	H	H
quercetin	OH	OH	H	H
vitexin	H	H	H	glucose
orientin	OH	H	H	glucose
saponaretin	H	H	glucose	H
homo-orientin	OH	H	glucose	H
vicenin	H	H	glycoside	glycoside
lucenin	OH	H	glycoside	glycoside

B

Aglycones	Substituents R	R′
cyanidin	OH	H
petunidin	OCH_3	OH

FIGURE 6.6.2 Flavonoid nucleus, numbering system, and substitutions (R) of the (A) flavone, flavonol, and glycoflavones and (B) anthocyanidins detected in taxa A–U.

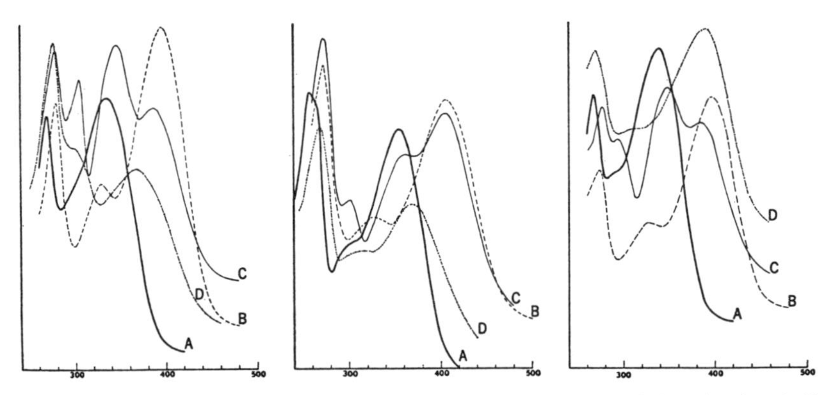

FIGURE 6.6.3 Ultraviolet absorption spectra of compound 2 (vitexin; left), compound 4 (quercetin-3-monoglycoside; center), and compound 29 (apigenin-7-monoglycoside; right). Ordinate = relative absorbance. Abscissa = wavelength in millimicrons. A = spectrum in methanol. B = spectrum in methanol plus sodium ethoxide. C = spectrum in methanol plus aluminum chloride. D = spectrum in methanol plus sodium acetate.

TABLE 6.6.1.
Origin of Experimental Clones of Taxa A–U

Taxon A
Uruguay: 115†

Taxon B
Brazil: 70, 70A, 70B, 70C, 70D, 70E, 70F, 70G, 70H, 70I; California: 6631; Louisiana: 51, 94, 95; Massachusetts; SOH†; North Carolina: 64; Texas: 7, 21

Taxon C
California: 6601 min, 6616, 6727; Florida: 91, 99, 100, 102; Louisiana: 97†; Texas: 74

Taxon D
Texas: 40†

Taxon E
California: 6601 tri; Connecticut: LTY†; Germany: 7013; Nebraska: 6722 tri; Switzerland: 6624

Taxon F
Louisiana: 53, 79†, 96

Taxon G
North Carolina: 57; Texas: 20†, 7000

Taxon H
Florida: 104, 105†

Taxon I
Illinois: 83; Texas: 22, 32, 36, 76†, 87

Taxon J
California: 6592; Florida: 26; Illinois: 62; Mississippi: 29; Nebraska: 6722w; North Carolina: 63†

Taxon K
California: 6613; Connecticut: SPY; England: 7017; Florida: 25, 101; Italy: 7010; Kansas: SPS; Louisiana: 24, 88, 98, 7003†; Mississippi: 30; Netherlands: 108, 112, 113; New Jersey: 6581; North Carolina: 58; Texas: 10, 23, 73, 86

Taxon L
California: 6566, 6583, 6602, 6745, 6751; Colorado: 41, 42; Germany: G1, G3†; Illinois: 46, 48; Italy: 6861, 7007; Manitoba: 6853; Montana: 6573; Nairobi: 89; Netherlands: 109; Oregon: 47, 6734; South Dakota: 38; Washington: 6735, 6736

Taxon M
California: 6609, 6612, 6746† (98-H, 99-B-1, 1073, 4130, 5104, 5280, 7130)[a], (L-1.4.3.1, L-1.4.3.2, L-1.4.3.3, L-1.4.3.4, L-1.4.3.5, L-1.4.3.6, L- 1.4.3.7, L-1.4.3.8, L-1.4.3.9, L-1.4.3.10, L-1.4.3.11, L-1.4.3.12, L-1.4.3.13)[b]; Florida: 27; Georgia: 28; Guatemala: 107; India: LPM; Nigeria: 77; South Carolina: 7006; Texas: 7001

Taxon N
Louisiana: 7002; Mississippi: 31; North Carolina: 56; Texas: 1, 18†

Taxon O
Brazil: 72†

Taxon P
India: WMI†

Taxon Q

Germany: 7014†, India: WAR; Netherlands: 111

Taxon R

Puerto Rico: 81†

Taxon S

British Columbia: 34-62; California: 44, 6582, 6587, 6591; Connecticut: LMY; Denmark: 59, 6579; England: 60, 7016; Illinois: 11; Italy: 7008, 7009; Kansas: LMS; Louisiana: 7004; Massachusetts: 49; Minnesota: 6; Missouri: 5†; Nebraska: 4; New Jersey: 6580; New Zealand: 7015; Pennsylvania: 6742; Saskatchewan: 8-62; Switzerland: 6578, 6623, 6625, 6626, 7011, 7019; Turkey: 7018; Vermont: 50; Washington: 6568, 6570, 10-62

Taxon T

California: 6584, 6589, 6600, 6711, 6726, 6744, 6747, 6863; Florida: 7005; Guatemala: 6717; Nebraska: 6752; Texas: 9†; Washington: 6737

Taxon U

Indiana: 55†

---

† Key clone of each taxon.

[a] X-ray induced aberrants of clone 6746 (Posner 1962. *Plant and Cell Physiol.* 3: 275–284).

[b] Progeny of selfed clone 6746 (Hillman, unpubl.).

TABLE 6.6.2.
Distribution of Flavonoids in Taxa A–U

	Glycoflavones												Antho-cyanins				Flavonols															Flavones										Unknown					
Taxa	1	2	7	15	16	18	20	23	24	25	26	27	8	11	12	19	3	4	5	9	10	22	36	37	38	40	41	42	43	44	45	13	14	17	21	29	32	34	35	46	47	6	28	30	31	33	39
A	*	*	*											*	*		*	*	*	*	*																					*					
B	*		*	*	*	*	*	*	*				*			*				*		*												*	*												
C				*		*	*			*	*			*	*																																
D						*	*			*																						*		*													
E	*	*		*		*	*			*	*			*	*																																
F																						*	*	*																							
G																									*																						
H																	*					*	*	*																							
I																	*		*	*	*			*	*	*	*	*			*																
J	*	*		*			*	*																										*						*							
K	*	*		*	*									*	*																	*	*	*								*					
L	*	*		*				*						*	*																																
M							*			*		*																					*														
N	*			*			*																													*	*	*	*		*	*	*	*			
O																	*					*	*	*																							
P																	*		*	*	*				*	*	*	*																			*
Q	*			*			*	*																										*						*	*						
R	*	*		*	*									*	*																	*	*	*								*					
S	*	*	*	*	*	*	*	*																																							
T						*	*			*	*																																				
U																	*		*	*	*							*	*	*	*																

NOTE: Numbers heading columns refer to individual compounds detected.

TABLE 6.6.3.

Identity, Color Reactions on Paper Chromatograms, and Spectral Maxima of Flavonoids (Exclusive of Anthocyanins) in the Taxa A–U

Compound	Identity	Color Reactions on Paper Chromatograms			Absorption Maxima in mμ			
		UV	$+NH_3$	+Benedict's	MetOH	MetOH−NaOEt	MetOH−NaOAc	MetOH−$AlCl_3$
1	orientin	D	YG	D	349,294a,268,256	405,330,278a,268	401a,352,267,256	388,360,295,274
2	vitexin	D	dYG	Y	333,300inf,269	394,328,278	365,300,278	383,344,303,276
3	Q-3-monoglycoside	D	YG	D	357,305,265a,256	414,328,275	370,310inf,268	404,360,300,270
4	Q-3-monoglycoside	D	YG	D	355,298,265a,257	418,326,274	366,315inf,269	403,359,299,277
5	K-3-monoglycoside	D	dYG	D	350,300,265	400,324,274	368,303,271	396,348,304,275
6	Unknown	D	D	D	333,300inf,265	388,320a,272	390,330,278	397,336,304,275
7	isosaponarin	D	YG	YG	332,270	397,329,277	376,300,277	385,346,303,279
9	Q-aglycone	Y	lY	lO	369,299,257	unstable	unstable	425,363,303,266
10	K-aglycone	Y	lY	lO	362,316,273	unstable	unstable	418,352,304,269
13	L-7-monoglycoside	D	Y	D	347,266,254	395,300a,269	420,352,266,256	394,360,295,273
14	A-7-monoglycoside	D	Y	lY	336,267	385,350,305,268	390,343,267	380,341,298,276
15	homo-orientin	D	dYG	D	348,292a,270,257	408,328,272	366,328,272	387,362,298,277
16	lutonarin	D	YG	D	348,292,269,255	405,340,269	400a,354,268,258	399,356,298,278
17	L-aglycone	D	YG	D	346,290,266,254	410,315,274	410a,354,266,256	388,358,293,273,262
18	lucenin	D	YG	D	350,272,259	420,340inf,272	380,330,282	400,368,300,280
20	vicenin	D	YG	Y	333,272	399,331,284	380,330,282	378,346,303,278
21	L-7-diglycoside	D	lYG	D	349,266,256	406,266	410,349,266,256	391,358,294,274
22	Q-3-7-diglycoside	D	YG	D	352,298,265,256	410,315,275	358,296,265,258	398,356,298,271
23	saponaretin	D	YG	YG	334,270	398,330,278	354,298,276	380,347,303,279
24	acylated saponaretin	D	lYG	lYG	326,272	385,334,275	325,298,275	328,302,279
25	acylated vicenin	D	YG	D	334,300,273	385,276	382,325,280	386,336,301,280
26	acylated lucenin	D	YG	Y	324,274	375,282	383,313,282	383,328,304,280
27	acylated vicenin	D	YG	Y	325,300,273	370,334,287	378,308,281	380,332,305,281
28	Unknown	D	Y	Y	341,286	395,300,268	390,268,256	382,346,297,275

29	A-7-monoglycoside	D	D	Y	339,290,270	396,329,274	386,306,268	383,348,293,277
30	Unknown	D	YG	Y	333,271	384,306	393,341	380,338,265,278
31	Unknown	D	Y	lY	338,270	389,305	390,264,257	382,346,295,276
32	A-7-diglycoside	D	dYG	dY	338,270	400,330,279	392,310,277	378,346,296,278
33	Unknown	D	Y	Y	332,269	391,355,305	394,256	384,346,294,276
34	A-7-monoglycoside	D	lYG	YG	334,330a,268	394,328,275	384,268	382,343,293,278
35	A-7-monoglycoside	D	dYG	YG	336,290a,268	396,326,275	381,269	383,345,292,276
36	Q-3-7-tri(?)glycoside	D	YG	D	352,296,267,255	404,327,271	367,324,268	406,360,298,271
37	Q-3-monoglycoside	D	YG	YG	357,396,267,255	424,330,280	385,284,269,258	398,362,296,268
38	Q-3-7-tri(?)glycoside	D	dYG	D	356,265a,256	408,270	420,366,258	410,360,300,273
39	Unknown	D	YG	D	330,293,252	382,270	not taken	392,332,300,272
40	Q-3-monoglycoside	D	dYG	D	357,300,256	410,329,272	374,323,271	408,360,300,272
41	Q-3-7-diglycoside	D	D	D	352,266,255	379,304,272	361,316a,273	394,347,286,274
42	K-3-7-diglycoside	D	YG	lY	347,265	386,350,300,272	354,265	392,342,296,275
43	Q-3-diglycoside	D	YG	D	357,300,257	410,321,270	385,268	416,360a,300,272
44	K-3-7-tri(?)glycoside	D	D	Y	348,295,265	400,323,275	363,301,268	395,347,302,274
45	Q-3-7-tri(?)glycoside	D	Y	D	358,263a,255	411,273	404,362,258	404,361,300a,270
46	L-7-diglycoside	D	dYG	D	348,293,270,258	407,340,267	388,317,280	387,259,298,276
47	L-(?)-glycoside	D	YG	D	349,291,269,256	406,336,269	386,324,276	410,340a,300a,276

NOTE: Compound number refers to the chromatographic spots illustrated in figure 6.6.1. Color of spot refers to its appearence on the chromatogram when viewed in ultraviolet light (UV); UV and in ammonia vapor (+NH_3); and UV after spraying with Benedict's reagent (+Benedict's). D = dark, Y = yellow, G = green, O = orange, d = prefix denoting intense (dark), l = prefix denoting weak (light). Maxima refer to the positions of maximum absorption in millimicrons. A = apigenin, K = kaempferol, L = luteolin, Q = quercetin.

The Maple-Vail Book *Composition Services*

DATE: 04-15-94 00:46:47 JOB ID: t18773$$t5
TITLE & MSPP:

Case Study Exercise 7 (Compositae)

PROBLEM: Provide a classification for the three taxa (A–C), using any or all of the following categories: species, subspecies, variety, form.

MATERIALS AND METHODS: A total of 65 individuals from six populations of Taxon C, 101 individuals from nine populations of Taxon A, and 60 individuals from four populations of Taxon B were sampled and examined for electrophoretic variation (table 6.7.1). Fruit collections were obtained from individual plants in each population and were propagated in the University of Oklahoma Botany greenhouses. One plant propagated from each individual fruit collection served as the source of material used in the electrophoretic analysis. Young leaves taken from either seedlings or mature plants served as the best source of enzymes. Identifications of taxa were based on morphological characters delineated by Shinners (1958. *Spring Flora of the Dallas–Fort Worth Area, Texas*. Dallas: published by author) and Northington (1971. Ph.D. Diss. Austin: University of Texas; 1974. *Spec. Publ. Mus.* No. 6. Lubbock: Texas Tech University). Voucher collections from sampled populations were deposited at OKL.

The extracting buffer consisted of 0.1 M Tris HCl pH 7.5, 1 mM EDTA (tetrasodium salt), 10 mM $MgCl_2$, 10 mM KCl, 14 mM 2-mercaptoethanol, and solid polyvinylpyrrolidone (5–10 mg/ml) (Gottlieb. 1981. *Proc. Natl. Acad. U.S.A.* 78: 3726–3729). Plant extracts were centrifuged in 1.5 ml tubes and the supernatant was absorbed onto wicks of Whatman 17MM chromatography paper. Samples were loaded and enzymes resolved on 11% starch gels.

Eight enzyme systems representing 13 loci were resolved using three buffer systems. System I consisted of an electrode buffer of 0.180 M Tris and 0.004 M EDTA titrated to pH 8.6 with boric acid, and a gel buffer obtained from a 1:3 aqueous dilution of the electrode buffer (Soltis et al. 1983. *Amer. Fern J.* 73: 9–27). System II, modified from Soltis et al. (1983), had an electrode buffer of 0.065 M L-histidine free base titrated to pH 6.5 with citric acid monohydrate, and a gel buffer obtained from a 1:3 aqueous dilution of the electrode buffer. System III consisted of an electrode buffer of 0.038 M monohydrate lithium hydroxide and 0.192 M boric acid titrated to pH 8.2, and a gel buffer obtained from a 1:9 dilution of the electrode buffer and 0.051 M Tris 7–9 and 0.007 M citric acid, pH 8.4.

System I was used to resolve aldolase (ALD), leucine aminopeptidase

(LAP), phosphoglucoisomerase (PGI), phosphoglucomutase (PGM), and malic enzyme (ME). System II was used to resolve aconitase (ACN) and 6-phosphoglucose dehydrogenase (6PGD). System III was employed to resolve triosephosphate isomerase (TPI). The staining methods for enzyme assays and agarose overlays (ALD, PGI, PGM, and TPI) followed Soltis et al. (1983).

Loci were numbered consecutively beginning with the most anodal when more than one isozyme was present. Similarly, alleles were assigned letters with the most anodal form designated a, and progressively slower forms designated b, c, and so forth.

Allele frequencies were determined for each population. The genetic basis of the observed patterns was inferred from several lines of evidence: patterns of variation among species with allelic differences, conformance to Hardy-Weinberg ratios in polymorphic taxa, and conformance to published patterns of enzyme expression and the basic number of loci expected in the absence of gene duplication or polyploidy (Gottlieb. 1982. *Science* 216: 373–380). Values for Nei's genetic diversity statistics—genetic identity, genetic distance, total gene diversity, diversity within and among populations, and differentiation among populations (Nei. 1972. *Amer. Nat.* 106: 283–292; 1973. *Proc. Natl. Acad. U.S.A.* 70: 3321–3323; 1978. *Genetics* 89: 583–590) were obtained for each pairwise comparison of populations and taxa using the GENESTAT program (Whitkus. 1988. *Pl. Genet. Newsletter* 4: 10). A UPGMA dendrogram based on genetic distance values was produced using the TAXON subroutine of the NTSYS program. Manual calculations include proportion of polymorphic loci, mean number of alleles per polymorphic locus, observed heterozygous loci per individual, expected heterozygosity, fixation indices, chi-square values, and number of unique alleles per taxon. A polymorphic locus was defined on the basis of the most common allele having a frequency of 0.99 or less. Mean fixation index values (F_T) were obtained from all polymorphic loci in each population.

DATA: figure 6.7.1, tables 6.7.1 through 6.7.5.

REFERENCE: Petersen, K. A., W. J. Elisens, and J. R. Estes. 1990. *Syst. Bot.* 15: 534–543.

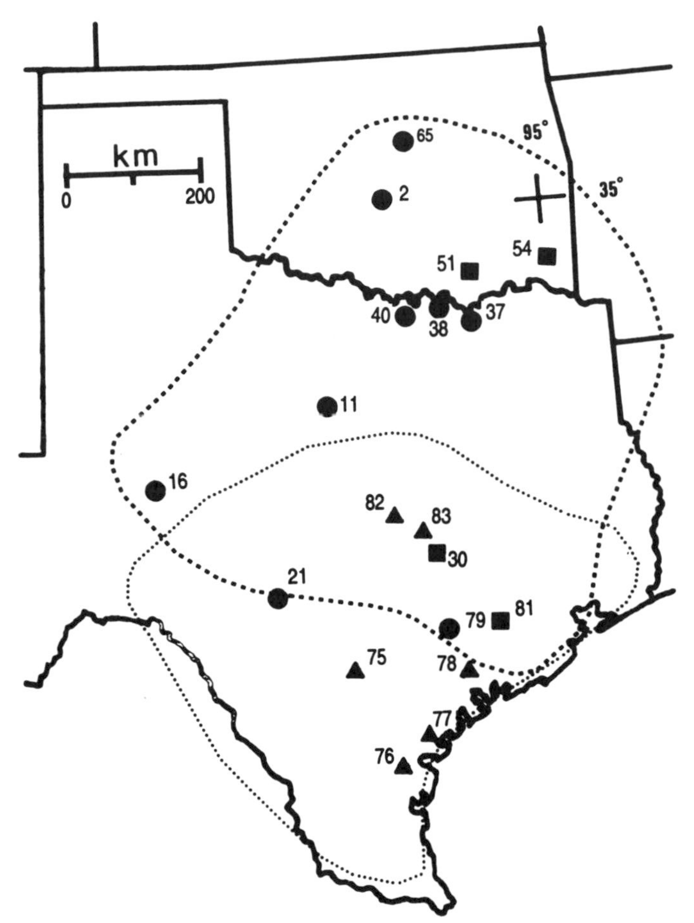

FIGURE 6.7.1 Geographic distribution and collection localities for Taxon A (circles), Taxon B (squares), and Taxon C (triangles). Range limits also shown for A (line of large dots) and C (line of small dots).

TABLE 6.7.1.
Collection Numbers and Sites for 19 Populations of Taxa A—C Sampled for Electrophoretic Study

Taxon A

Oklahoma: Cleveland Co., Norman, Petersen 2 (16); Payne Co., Stillwater, Petersen 65 (6). **Texas:** Colorado Co., Garwood, Petersen 79 (6); Delta Co., Enloe, Petersen 37 (9); Fannin Co., Honey Grove, Petersen 38 (5); Gillespie Co., Fredricksburg, Petersen 21 (18); Grayson Co., Sherman, Petersen 40 (16); Palo Pinto Co., Mineral Wells, Petersen 11 (16); Tom Greene Co., San Angelo, Petersen 16 (9).

Taxon B

Oklahoma: McCurtain Co., Hochatown State Park, Petersen 51 (6); Sequoyah Co., Vian, Petersen 54 (19). **Texas:** Brazos Co., Bryan, Petersen 30 (24); Waller Co., Prairie View, Petersen 81 (11).

Taxon C

Texas: Aransas Co., Rockport, Petersen 77 (11); Atascosa Co., Fashing, Petersen 75 (7); Brazos Co., Bryan, Petersen 83 (15); Grimes Co., Navasota, Petersen 82 (10); Jackson Co., La Ward, Petersen 78 (16); San Patricio Co., Sinton, Petersen 76 (6).

NOTE: Number of individuals examined is in parentheses.

TABLE 6.7.2.
Summary of Allele Frequency Data for 11 Polymorphic Loci in Taxa A–C

Locus	Allele	Taxon		
		A	B	C
ACN-1	a	0.005	0.000	0.000
	b	0.009	0.000	0.017
	c	0.981	1.000	0.983
	d	0.006	0.000	0.000
ACN-2	a	0.016	0.000	0.017
	b	0.974	0.020	0.983
	c	0.010	0.897	0.000
	d	0.000	0.083	0.000
ALD-1	a	0.048	0.000	0.052
	b	0.937	1.000	0.921
	c	0.015	0.000	0.027
LAP-1	a	0.145	0.000	0.000
	b	0.740	0.000	0.336
	c	0.071	0.965	0.664
	d	0.044	0.035	0.000
ME-1	a	0.010	0.000	0.000
	b	0.758	0.000	0.908
	c	0.232	1.000	0.092
PGI-1	a	0.112	0.000	0.044
	b	0.517	0.018	0.545
	c	0.370	0.549	0.374
	d	0.000	0.433	0.037
PGI-2	a	0.000	0.000	0.008
	b	1.000	0.036	0.979
	c	0.000	0.948	0.014
	d	0.000	0.016	0.000
PGM-1	a	0.105	0.153	0.066
	b	0.895	0.847	0.934
PGM-2	a	0.011	0.000	0.000
	b	0.989	0.983	0.972
	c	0.000	0.016	0.028
6PGD-1	a	0.110	0.000	0.031
	b	0.306	0.000	0.009
	c	0.544	1.000	0.917
	d	0.030	0.000	0.017
	e	0.005	0.000	0.025
	f	0.005	0.000	0.000
6PGD-2	a	0.074	0.000	0.000
	b	0.820	0.037	0.942
	c	0.105	0.016	0.000
	d	0.000	0.911	0.058
	e	0.000	0.035	0.000

TABLE 6.7.3.
Genetic Variation Within Populations and Taxa A–C

Taxon	Collection Number	P	k_p	k	H_o	H_e	F_T	Unique Alleles
Within Populations:								
Taxon A	16	0.308	2.250	1.357	0.034	0.128	0.567	0
	21	0.385	2.250	1.357	0.009	0.090	0.902†	1
	37	0.615	2.750	2.000	0.082	0.224	0.581	1
	38	0.231	2.330	1.286	0.046	0.108	0.475	1
	40	0.615	2.620	1.929	0.106	0.130	0.148	1
	65	0.385	2.200	1.429	0.051	0.128	0.310	0
	11	0.462	2.330	1.571	0.088	0.175	0.494†	0
	2	0.385	2.800	1.643	0.060	0.167	0.791†††	0
	79	0.231	2.000	1.214	0.064	0.103	0.253	0
Taxon B	81	0.385	1.800	1.357	0.007	0.084	0.926	0
	30	0.308	2.250	1.357	0.042	0.058	0.870	0
	51	0.154	2.000	1.143	0.051	0.038	−0.215	0
	54	0.231	2.330	1.286	0.020	0.045	0.357	1
Taxon C	75	0.308	2.250	1.357	0.057	0.114	0.474	0
	76	0.538	2.000	1.429	0.064	0.141	0.532	0
	77	0.538	2.000	1.500	0.035	0.084	0.570	0
	78	0.615	2.250	1.786	0.091	0.120	0.270	1
	82	0.077	2.000	1.071	0.031	0.031	0.170	0
	83	0.385	2.000	1.500	0.041	0.062	0.315	0
Within taxa:								
Taxon A		0.402	2.392	1.532	0.061	0.141	0.551	7
Taxon B		0.249	2.095	1.286	0.030	0.070	0.556	3
Taxon C		0.410	2.083	1.441	0.053	0.082	0.389	1

NOTE: Table heads refer to proportion of polymorphic loci (P), mean number of alleles per polymorphic locus (k_p), mean number of alleles per locus (k), average observed heterozygosity (H_o), average expected heterozygosity (H_e), mean fixation index (F_T), and unique alleles. Collection numbers refer to populations listed in table 6.7.1.

†$p > 0.05$

†††$p > 0.001$

TABLE 6.7.4.
Mean Values for Nei's Genetic Diversity Statistics Within Individual Polymorphic Loci and Pooled Polymorphic Loci Within Taxa A–C

	H_t	H_s	D_{st}	G_{st}
Within loci:				
ACN-1	0.022	0.021	0.001	0.009
ACN-2	0.363	0.065	0.298	0.820
ALD-1	0.113	0.098	0.015	0.130
LAP-1	0.566	0.268	0.298	0.526
ME-1	0.468	0.155	0.313	0.668
PGI-1	0.640	0.224	0.415	0.649
PGI-2	0.331	0.030	0.301	0.907
PGM-1	0.214	0.184	0.030	0.142
PGM-2	0.031	0.031	0.001	0.017
6PGD-1	0.421	0.273	0.147	0.350
6PGD-2	0.468	0.158	0.310	0.662
Within taxa:				
Taxon A	0.206	0.152	0.054	0.263
Taxon B	0.111	0.058	0.053	0.473
Taxon C	0.155	0.101	0.054	0.346

NOTE: Table heads refer to total gene diversity (H_t), gene diversity within populations (H_s), gene diversity among populations (D_{st}), and differentiation among populations ($G_{st} = D_{st}/H_t$).

TABLE 6.7.5.
Summary Values for Nei's Genetic Identity (I) and Genetic Distance (D) Coefficients for Within and Between Taxa A–C

	I		D	
	Mean	Range	Mean	Range
Within Taxa:				
Taxon A	0.928	0.842–0.994	0.076	0.006–0.172
Taxon B	0.926	0.878–0.990	0.078	0.010–0.130
Taxon C	0.929	0.866–0.993	0.074	0.007–0.144
Between taxa:				
Taxon B × A	0.623	0.455–0.724	0.474	0.323–0.786
Taxon B × C	0.676	0.579–0.794	0.391	0.231–0.547
Taxon A × C	0.900	0.745–0.988	0.107	0.012–0.295

Case Study Exercise 8 (Linaceae)

PROBLEM: Provide a classification into informal groups for the taxa A–H.

MATERIALS AND METHODS: *Plant Material:* Plants were grown from seed in John Innes No. 1 potting compost in the glasshouse with no supplementary lighting. Plants were harvested after 6–8 wk, and the leaves and apical meristems were used for all nucleic acid extractions. The taxa used, and their karyotypic characteristics, are listed in table 6.8.1. Seed for all of these came from the John Innes Institute collections.

Isolation of DNA: Chloroplasts of Taxon H were isolated and the DNA extracted using the method of Kolodner and Tewari (1975. *Nature* 256: 708–711). Total DNA from the other taxa was isolated as described by Cullis (1979. *Heredity* 42: 237–246).

Restriction endonuclease digestion of DNA: Total or cpDNAs were digested at 37°C with the restriction endonucleases SalP1, subsequently denoted by Pst1 (gift of Dr. P. B. Goldsbrough), BamH1, Sst1, and SalG1, subsequently denoted by Sal1 (Miles Laboratories Ltd.). The BamH1, Sst1, and Sal1 digestions were carried out as suggested by the supplier. The Pst1 digestions were carried out in the buffer suggested by Carter et al. (1980. *Nucleic Acids Res.* 8: 4943–4954).

Agarose gel electrophoresis: DNA fragments were separated by gel electrophoresis and transferred to nitrocellulose filters (Schleicher and Schuell BA85) as previously described (Goldsbrough and Cullis. 1981. *Nucleic Acids Res.* 9: 1301–1309). All gels included marker tracks consisting of the partial double digest fragments of bacteriophage = DNA cleaved with EcoR1 and HindIII.

Radiolabeling of DNA probes: Probes for hybridization were nick-translated in the presence of α-^{32}P-dCTP or α-^{32}P-dATP (Maniatis, Jeffrey, and Kleid. 1975. *Proc. Natl. Acad. Sci. U.S.A.* 72: 1184–1188).

Characteristics of DNA probes: The cloned chloroplast DNAs from Taxon H used in this study are listed in table 6.8.2.

DATA: figures 6.8.1 through 6.8.4; tables 6.8.1 through 6.8.4.

REFERENCE: Coates, D. and C. A. Cullis. 1987. *Amer. J. Bot.* 74: 260–268.

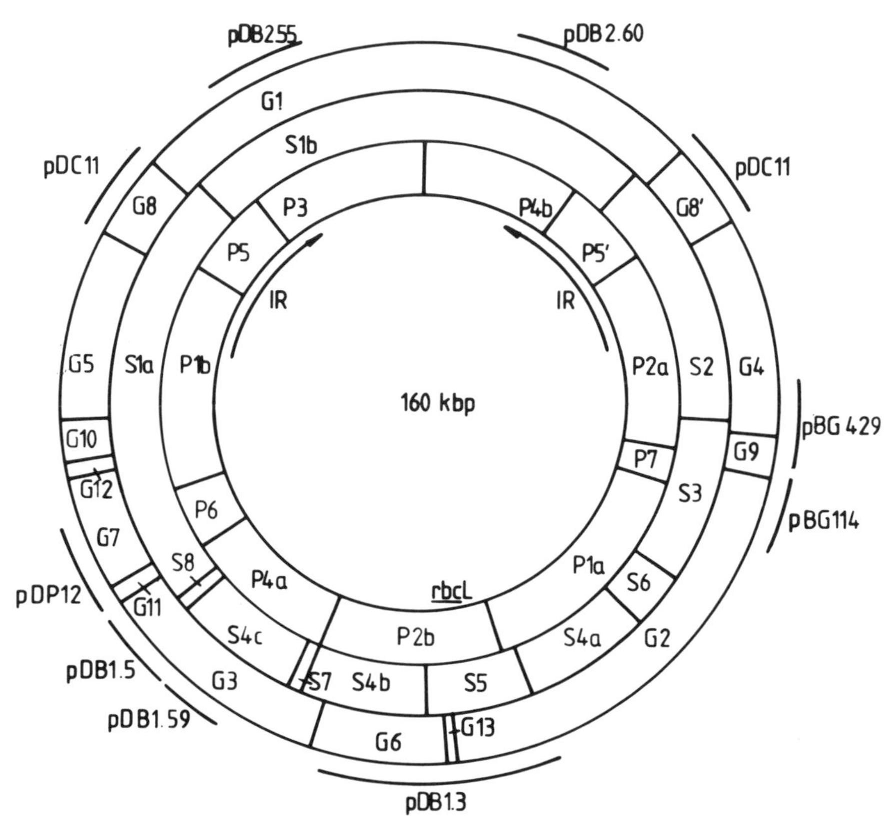

FIGURE 6.8.1 Restriction map of Taxon H cpDNA. P = Pst1 fragments; S = Sst1 fragments; G = Sal1 fragments; IR = inverted repeat; *rbc*L = gene coding for the large subunit of ribulose 1,5-bisphosphate carboxylase. The outer circle shows the positions and names of the various recombinant DNA clones (see table 6.8.1).

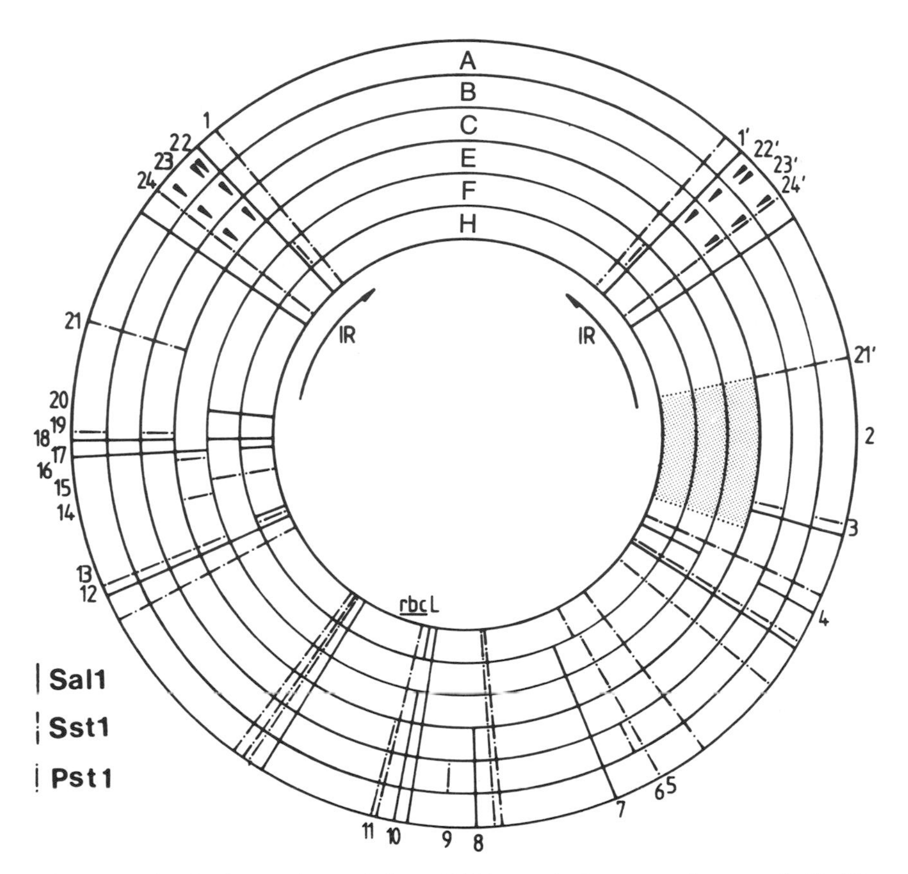

FIGURE 6.8.2 Comparison of the cpDNA restriction maps for taxa A–H. IR = inverted repeat; *rbc*L = gene coding for the large subunit of ribulose 1,5-bisphosphate carboxylase. Narrow triangles = sites of short insertion/deletions relative to Taxon H cpDNA. The shaded area represents a large deletion.

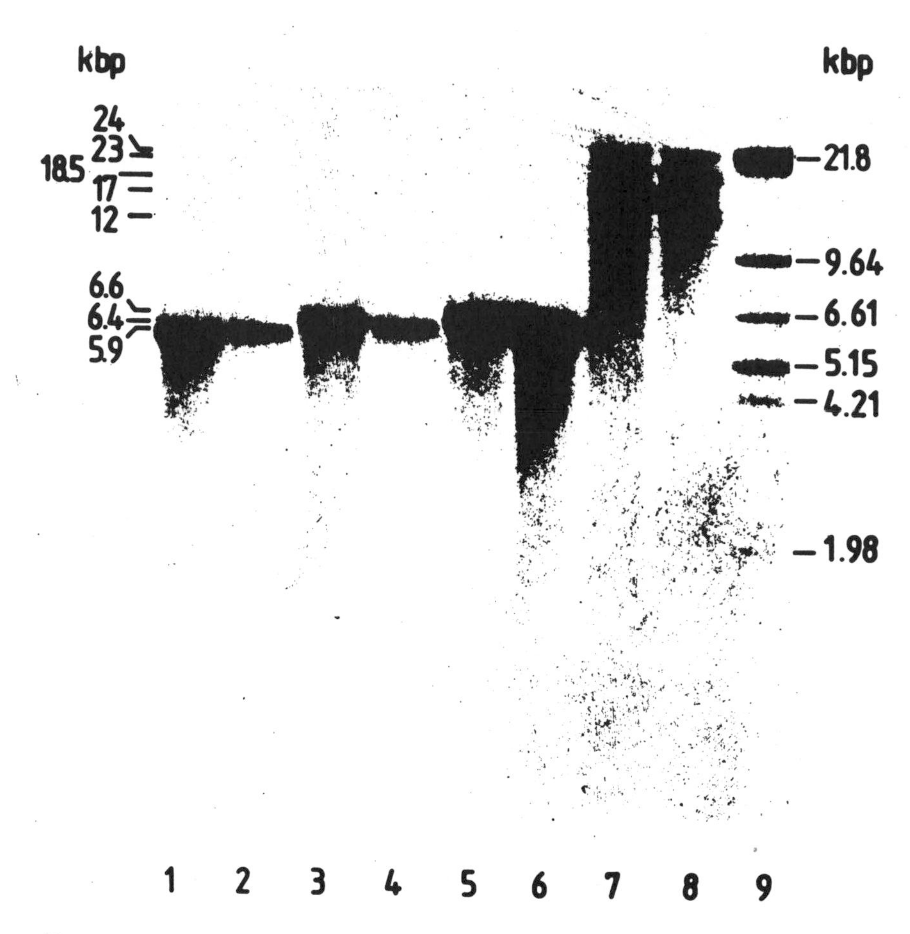

FIGURE 6.8.3 Southern transfer of taxa A–H cpDNAs cleaved with Sal1 and hybridized with pDC11. 1 = Taxon H; 2 = Taxon E; 3 = Taxon C; 4 = Taxon A; 5 = Taxon D; 6 = Taxon F; 7 & 8 = two other more distantly related taxa outside group A–H; 9 = partial digest of DNA with EcoR1 and HindIII.

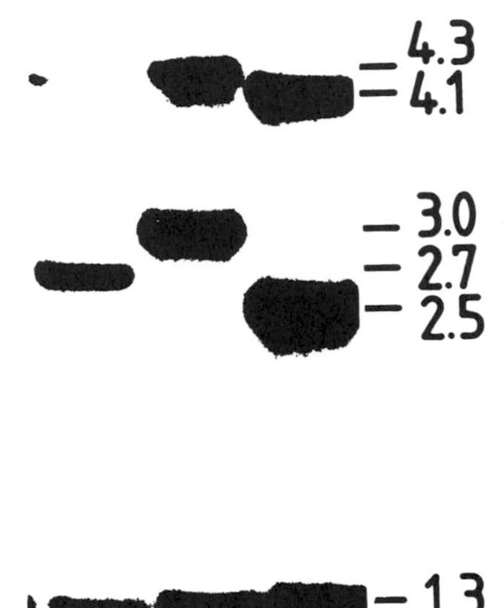

FIGURE 6.8.4 Southern transfer of taxa A–H cpDNAs cleaved with BamH1 and hybridized with pDC11. 1 = Taxon A; 2 = Taxon D; 3 = Taxon H. Note the size variations among the three taxa. The 4.1 kb band in Taxon H cpDNA has increased to 4.3 kb in the other two taxa. The 2.5 kb band is 2.7 kb in Taxon A and 3 kb in Taxon D. This has been interpreted as an initial 200 bp insertion/deletion separating some taxa, followed by a 300 bp insertion separating others.

TABLE 6.8.1.
Karyotypes of Taxa A–H

Taxon	2*n*	Chromosome Size
A	18	medium
B	18	medium
C, D	18	medium
E	28	large
F	16	medium
G, H	30	small

SOURCES: Ray. 1944. *Amer. J. Bot.* 31: 241–248; Ockendon and Walters. 1968. In Tutin et al., *Flora Europaea* 2: 206–211.

TABLE 6.8.2.
Recombinant Probes Isolated from Taxon H cpDNA Used to Compare the cpDNA Restriction Maps of the Other Taxa

Recombinant Probe	Insert Site and Size
pBG114	BamH1, 4.9 kbp
pBG429	BamH1, 6.1 kbp
pDC11	Sal1, 5.9 kbp (G8)
Includes the 16S rDNA and part of the 23S rDNA:	
pDP12	Pst1, 6.1 kbp (P6)
pDB1.3	BamH1, 13.5 kbp
Includes rbcL:	
pDB1.5	BamH1, 4.0 kbp
pDB1.59	BamH1, 4.5 kbp
pDB2.55	BamH1, 6.6 kbp[a]
pDB2.60	BamH1, 6.0 kbp[a]

NOTE: Designations in parentheses refer to the specific restriction fragments shown in figure 6.8.1.

[a] Spans the inverted repeat/short single copy sequence boundaries.

TABLE 6.8.3.
Comparative Differences Between Restriction Maps of Various cpDNAs

		Taxon					
Site #	Variation	A	B	CD	E	F	GH
1	P					+	
2	I	+	+	+			
3	P	+		+	−	−	−
4	G	+	+			+	+
5	P	+	+	+	+	+	
6	S		+				+
7	G	+	+	+	+	+	
8	G	+	+	+			
9	P		+				
10	G	+	+	+	+		+
11	P	+	+	+			
12	P	+	+	+	+		+
13	G						+
14	P				+		
15	P					+	+
16	P				+		
17	G	+	+	+	+		+
18	G	+	+	+		+	+
19	P	+		+			
20	G					+	+
21	S	+	+	+			
22	I	+	+	+			
23	I	+					
24	I	+	+	+			

NOTE: Site # refers to the position of the variation shown in figure 6.8.2. G = SalI; P = PstI; S = SstI; I = insertion relative to Taxon H; − = not applicable; + = site present.

TABLE 6.8.4.
Analysis of Restriction Fragments and Sequence Divergence of Plastid Genomes in Taxa A–H

	G, H	F	E	C, D	A	B
Estimated percentage of base pair substitutions per base pair (P*):						
G, H	0	6.5	8.07	9.13	9.82	11.95
F		0	8.47	9.54	12.56	13.82
E			0	9.26	8.97	10.0
C, D				0	1.9	5.07
A					0	5.53
B						0
Percentage of shared fragments (F**):						
G, H	100	62	68	56	58	49
F		100	60	59	58	55
E			100	47	57	44
C, D				100	89	72
A					100	74
B						100

SOURCE: Nei and Li 1979. *Proc. Natl. Acad. Sci. U.S.A.* 76: 5269–5273.
*$P = -(3/2)\ln((4F^{1/2r} - 1)/3)$
**$F = 2n_{xy}/n_x + n_y$

Case Study Exercise 9 (Onagraceae)

PROBLEM: Provide a classification for the nine taxa (A–I), using any or all of the following categories: species, subspecies, variety, form.

MATERIALS AND METHODS: Observations of meiotic chromosomes were made from 61 populations widely distributed over the ranges of the included taxa. Plants from 34 populations, including material for each of the nine taxa, were tested for self-compatibility. For the estimates of pollen grain stainability, 300 grains minimum were counted after having been stained with lactophenol cotton blue. For the morphological data, many plants of each taxon were analyzed. These many vouchers will be cited in a forthcoming paper on the taxonomy of this group.

DATA: figures 6.9.1 through 6.9.4; tables 6.9.1 through 6.9.6.

REFERENCE: Klein, W. M. 1970. *Evolution* 24: 578–597.

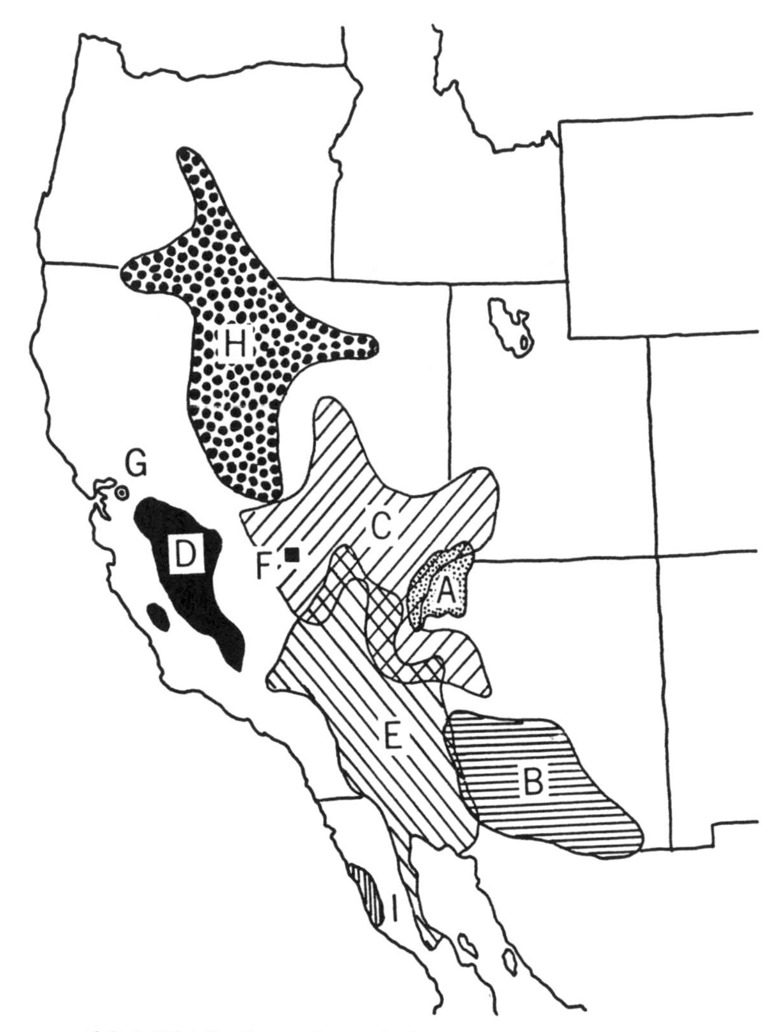

FIGURE 6.9.1 Distributions of taxa A–I.

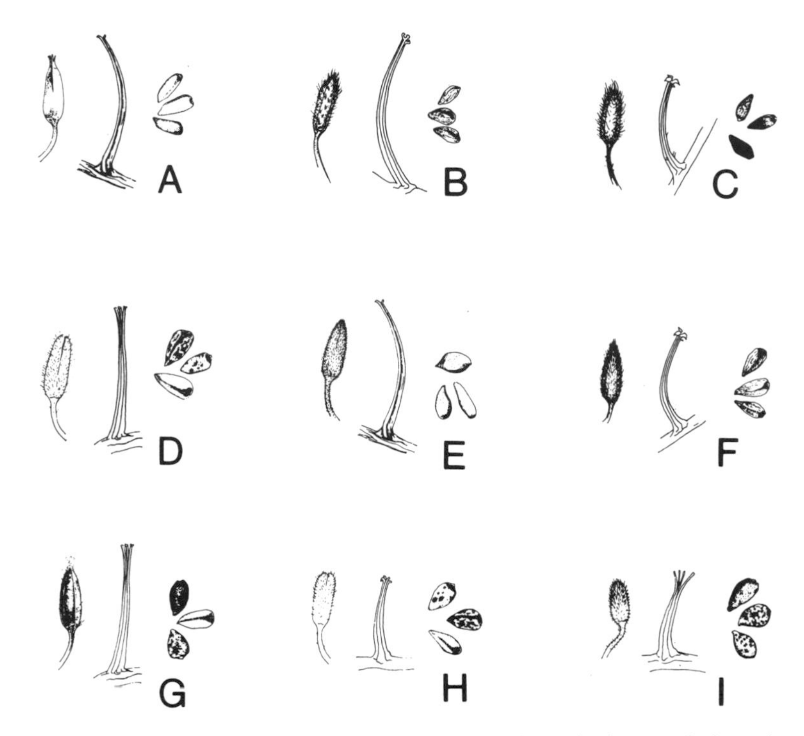

FIGURE 6.9.2 Representative bud, capsule, and seed characteristics of taxa A–I. Bud and capsules are ×0.5 and seeds are ×3.

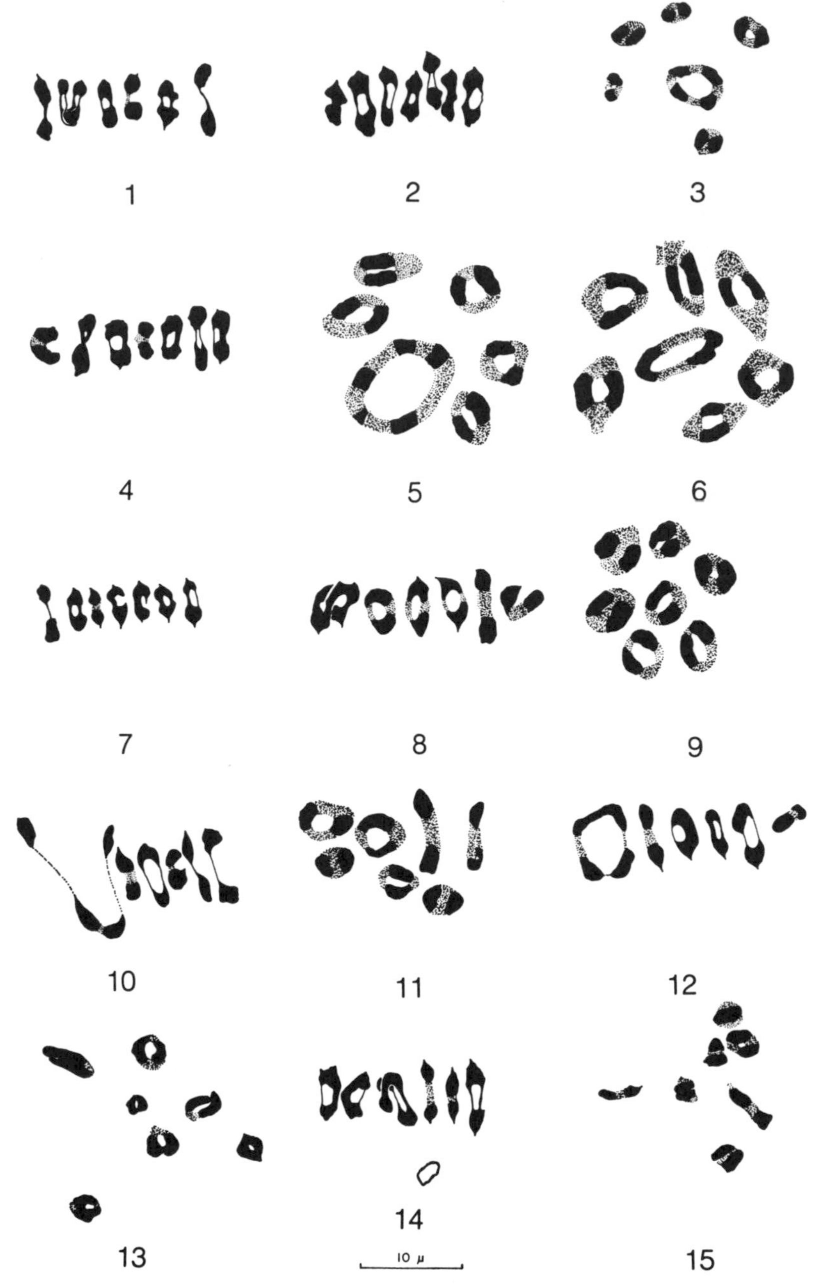

FIGURE 6.9.3 Chromosome pairing of intertaxon hybrids. Camera lucida drawings of diakinesis and metaphase I. 1, Taxon B × Taxon C, r4 & 5II; 2, C × F, 7II; 3, C × F, r4 & 5II; 4, A × D, 7II; 5, A × D, r4 (extra dark staining segment in ring) & 5II; 6, A × E, 7II; 7, D × E, 7II; 8, D × E, r4 & 5II; 9, D × G, 7II; 10, D × G, ch4 & 5II; 11, E × G, 7II; 12, E × G, r4 & 5II; 13, E × H, 7II; 14, E × H, r4 & 5II; 15, G × H, 7II. r = ring; ch = chain; I = univalent; II = bivalent.

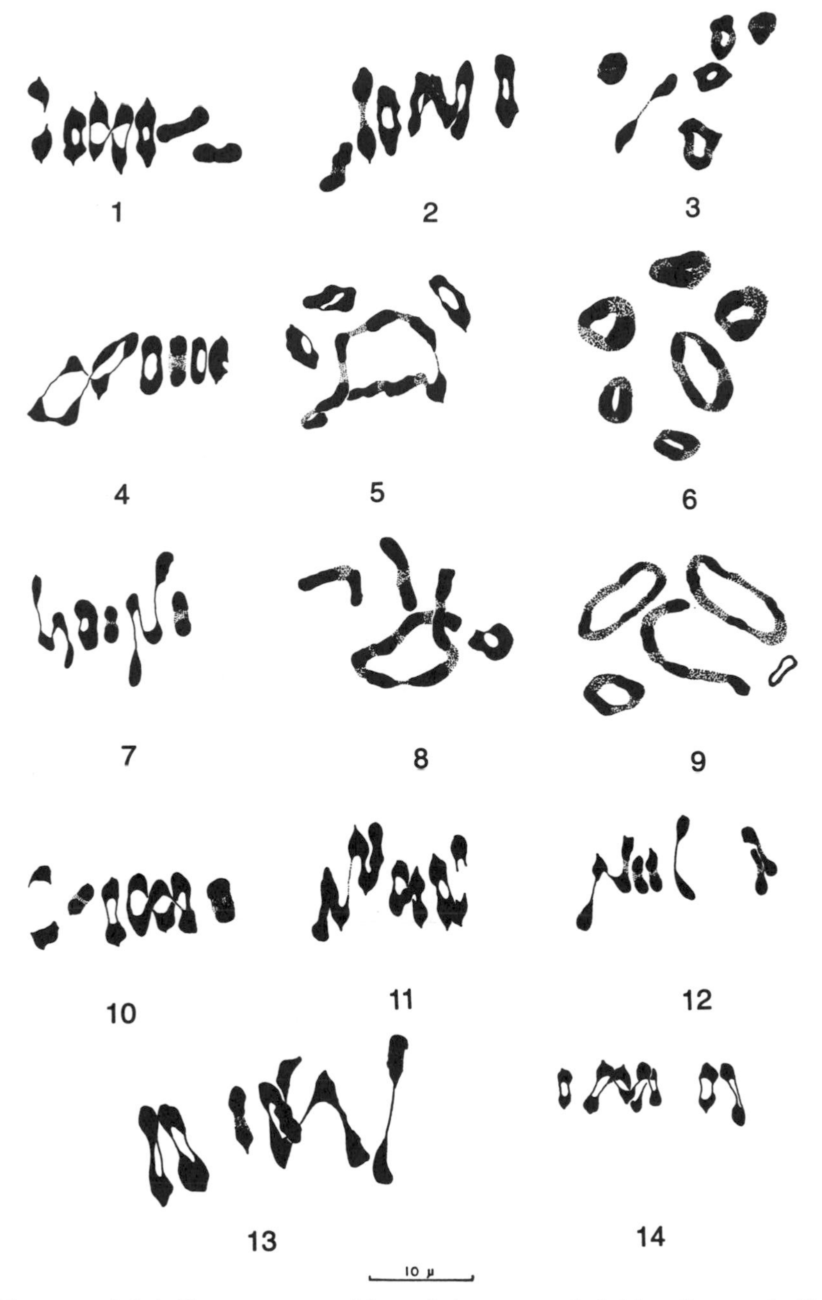

FIGURE 6.9.4 Chromosome pairing of intertaxon hybrids. Camera lucida drawings of diakinesis and metaphase I. 1, B × D, r4 & 5II; 2, B × D, r6 & 4II; 3, B × G, r4 & 5II; 4, B × G, r6 × 4II; 5, B × G, r8 & 3II; 6, C × E, r4 & 5II; 7, C × E, 2ch4 & 3II; 8, C × G, ch8 & 3II; 9, C × H, 2r4 & ch4 & II & I; 10, F × D, r6 & 4II; 11, F × D, ch6 & r4 & 2II; 12, F × G, 2ch4 & 3II; 13, F × G, ch6 & 4II; 14, I × G, r8 & 3II.

TABLE 6.9.1.
Comparative Morphology of Taxa A–I

Taxon	Duration	Growth Habit	Leaf Margin	Mature Buds	Flower (diam)	Capsule (Length)	Seeds
A	annual	1–5 dm tall; woody taproot	entire, dentate to lobed	2–3 cm long; lanceolate; free-ending sepal tips; strigose/glabrate	3–7 cm	3–8 cm	smooth surface; buff–lightly mottled
B	annual	1–3.5 dm tall; taproot	pinnatifid	1.9–2.5 cm long; prominent purple spots; sparingly to densely pilose	4.5–5.5 cm	5–8 cm	minutely ridged surface; buff–darkly mottled
C	perennial	1–3 dm tall; spreading underground stems	entire, to deeply incised	1.5–3 cm long, pilose	4–6 cm	2–8 cm	minutely ridged surface; olive brown, black
D	annual –short-lived perennial	2–6 dm tall; woody taproot	entire to runcinate; glabrous and bright green color	2–2.5 cm long, oblong–elliptical; rounded tip; pilose	5–8 cm	4–6 cm	smooth surface; buff–heavily mottled
E	annual	1–5 dm tall; woody taproot	entire–dentate to lobed	2–3 cm long; lanceolate; acute tip; velutinous/stigose	3–7.5 cm	3–8 cm	smooth surface; buff–lightly mottled
F	perennial	1–4.5 dm tall; spreading underground stems	entire to broadly dentate	1.5–3 cm long; densely pilose	5–6 cm	3–7 cm	minutely ridged; buff–darkly mottled
G	perennial	5–8 dm tall; fleshy roots	pinnatifid, undulating; densely strigose and gray green color	2.5–3 cm long; lanceolate; free-ending sepal tips; sparsely pilose/glabrous	6–8 cm	4–6 cm	smooth surface; buff–mottled black
H	annual	1–4 dm tall; woody taproot	divided near base of blade, pinnatifid	1.5–2.3 cm long; oblong–obovate; rounded, grooved at tips, pilose/glabrate	4–6 cm	2.2–3.5 cm	smooth surface; buff–heavily mottled–black
I	annual	1.5 dm tall; woody taproot	pinnately divided in lower half	1.0–1.8 cm long; oblong; rounded tip; densely pilose	3–5 cm	2–4 cm	smooth surface; olive brown–heavily mottled

TABLE 6.9.2.
Frequency of Meiotic Configurations in Natural Populations of Taxa A–I

Taxon	Populations Sampled	Plants Analyzed	Percentage of Various Meiotic Configurations		
			7II	1r4 + 5II	2r4 + 3II
A	10	21	95	5	
B	1	6	34	50	16
C	17	36	67	28	5
D	2	2	50	50	
E	23	60	82	18	
F	1	4	75	25	
G	1	2	50	50	
H	3	9	78	22	
I	3	14	100		

TABLE 6.9.3.
Meiotic Configurations of Plants from Natural Populations

Taxon A. Nevada. Clark County: Boulder City, Klein 968 (7II), 970 (7II), 971 (r4 + 5II), 973 (7II), Munz 22988 (7II); Junction of the Overton and Valley of Fire roads, Klein 993 (7II), 995 (7II); Meadow Valley Wash S of Rox, Klein 1003 (7II); Meadow Valley Wash 11 mi. S of Carp, Klein 1009 (7II), 1012 (7II); 13 mi. NE of Glendale, Klein 1019 (7II), 1021 (7II); 2.8 mi. N of Logandale, Klein 1022 (7II), 1025 (7II), 1026 (7II), 1027 (7II), 1029 (7II), 1032 (7II + 3B); 9 mi. NE of Dry Lake, Gregory 9 (7II); 6 mi. S of Overton, Gregory 17 (7II); 1 mi. E of Glendale, Gregory 12 (7II); 4 mi. E of U.S. Hwy. 93 on road to Valley of Fire, Gregory 382 (7II).

Taxon B. Arizona. Yuma County: 18 mi. NW of Hope along Ariz. Hwy. 72, Klein 954 (r4 + 5II), 955 (2r4 + 3II), 956 (r4 + 5II), 957 (7II); Grant 10074 (7II); Prop. RSABG (r4 + 5II).

Taxon C. Arizona. Mohave County: 10 mi. E of Kingman along U.S. Hwy. 66, Munz & Gregory 23527 (7II, r4 + 5II); 9 mi. NW of Kingman on Boulder City road, Munz 22983 (7II). **California.** Inyo County: 5 mi. S of Oasis, Klein 339 (7II). San Bernardino County: Cedar Canyon, Mid Hills between the New York and Providence Mts., Klein 925 (7II), 926 (7II), 1416 (7II); Gregory 373 (7II); 12.5 mi. S of Kelso on Calif. Hwy. 66, Klein 1345 (r4 + 5II), 7 mi. S of Kelso on Calif. Hwy. 66, Transplant RSABG (r4 + 5II); 1.5 mi. N of Essex on Mitchell Cavern road, Klein 1359 (r4 + 5II), 1360 (7II); 11 mi. N of Goffs, Munz 23005 (7II). **Nevada.** Clark County: 1 mi. NE of Jean along U.S. Hwy. 91-466, Klein 979 (7II), 981 (7II), 982 (7II), 984 (7II), Munz 23001 (7II), 23002 (7II); 12.8 mi. N of Moapa Junction, Klein 1006 (r4 + 3II), 1007 (2r4 + 3II). Lander County: 2 mi. W of Austin, Klein 1069 (r4 + 5II), 1073 (r4 + 5II), 1074 (r4 + 5II). Lincoln County: White Clay Hills E of Panaca, Raven 11256 (7II[2]). Nye County: 21 mi. E of Warmsprings along U.S. Hwy. 6, Klein 1161 (7II), 1163 (7II), 1164 (r4 + 5II). **Utah.** Beaver County: 8.5 mi. W of Milford, Klein 1062 (7II), 1064 (7II); 23 mi. W of Milford, Klein 1060 (7II). Washington County: 1/2 mi. NE of Leeds, Klein 1049 (7II, r4 + 5II); Pine Valley, Klein 1052 (7II), 1054 (r4 + 5II).

Taxon D. California. Merced County: 20 Mi. W of Merced along Calif. Hwy. 140, Klein 443 (7II); north side of Merced River on U.S. Hwy. 99, RSABG Prop. No. 8625 (r4 + 5II).

Taxon E. Arizona. Yuma County: 2.5 mi. NW of Bouse, Klein, 960 (4r + 5II), 961 (7II), 962 (7II), Prop. RSABG (7II); 10 mi. E of Yuma on U.S. Hwy. 95, Munz 22965 (7II). **California.** Imperial County: 3.6 mi. S of Kane Spgs., Klein 61 (7II), 62 (7II); 4 mi. S of Desert Edge along U.S. Hwy. 80, Klein 74 (7II), 75 (7II). Los Angeles County: NE of Wilson School, Klein 289 (r4 + 5II). Riverside County: 4.5 mi. N of Ripley, Klein 147 (7II); Bermuda Dunes, Klein 797 (7II), 799 (r4 + 5II), 800 (r4 + 5II), 802 (7II), 805 (7II), 806 (r4 + 5II); across RR tracks from Thousand Palms, Gregory 360 (r4 + 5II); 3.5 mi. NW of Indio, Munz 22962 (7II). San Bernardino County: N of Twentynine Palms along the Amboy rd., Klein 240 (7II); 8 mi. N of Hinkley, Klein 808 (7II), 811 (7II), 812 (7II), 814 (7II), 815 (7II); 10 mi. N of Hinkley, Klein 819 (7II); 10 mi. SW of Cronese Valley Station along U.S. Hwy. 91-466, Klein 832 (7II), 833 (7II), 834 (7II), 854 (7II), 859 (7II); Cronese Valley Station, Klein 836 (7II), 838 (7II), 841 (r4 + 5II); Dunes 2 mi. S of Kelso, Klein 892 (7II), 896 (7II), 897 (7II), 898 (7II), 901 (r4 + 5II), 903 (7II), 904 (7II), 922 (r4 + 5II), 923 (7II); Essex along the N side of U.S. Hwy. 66, Klein 931 (7II); 1.5 mi. NW of Essex along the road to Mitchell Caverns, Klein 1361 (7II), 1362 (7II), 1363 (7II); 2.8 mi. W of Adelanto along the road to Palmdale, Klein 1336 (7II), 1337 (7II), 1338 3(7II); 20 mi. E of Twentynine Palms along the road to Rice, Klein 1368 (7II); Daggett, Transplant propagated at RSABG (7II[4]); 17 mi. E of Twentynine Palms, W end of Dale Dry Lake, Munz 23008 (7II + 1B, r4 + 5II); 32 mi. E of Twentynine Palms along the road to Rice, Klein 1466 (7II). **Mexico.** Sonora: Behind Bahía la Cholla near Puerto Penasco, Raven 14812 (r4 + 5II).

Taxon F. California. Inyo County. Sand dunes at the SE end of Eureka Valley, Klein 383 (7II), 384 (7II), RSABG Prop. No. 9063 (7II), Gregory 394 (r4 + 5II).

Taxon G. California. Contra Costa County: Antioch, Klein 440 (r4 + 5II), RSABG Prop. No. 9484 (7II).

Taxon H. Nevada. Ormsby County: 4 mi. E of Carson City on U.S. Hwy. 50, Klein 419 (7II); 5 mi. E of Carson City on U.S. Hwy. 50, Klein 1079 (7II), 1080 (7II), 1081 (r4 + 5II), 1082 (7II), 1085 (7II), 1086 (7II). Lyons County: 16 mi. NE of Carson City, Klein 425 (7II), 426 (r4 + 5II).

Taxon I. Mexico. Baja California: Sandhills along the coast of Bahía San Quintín about 2 mi. SE of the cannery, Klein 1202 (7II[12]); along the coast at Santa Maria beach 7 mi. S of San Quintín, Klein 1221 (7II); 1.5 mi. E of Punta Baja along the road to Rosario, Klein 1237 (7II).

NOTE: Associations of chromosomes were observed at diakinesis or metaphase I. The symbols should be read II = bivalent, r = translocation ring, and ch = chain indicating occurrence of translocation. If more than one plant was analyzed the number appears in brackets.

TABLE 6.9.4.
Compatibility Relationships of Plants from Natural Populations of Taxa A–I

Taxa	Self-Compatible	Self-Incompatible	Populations Tested
A		X	Boulder City; Carp; Logandale
B	X		Hope
C		X	Amboy; Austin; Cedar Canyon; Jean; Locke Station; Pine Valley
D	X		Merced; Merced River
E	X		Bermuda Dunes; Indio
		X	Barstow; Blythe; Bouse; Cronese; Daggett; Dale Dry Lake; Essex; Hinkley; Kelso; Las Vegas; Rice; Twentynine Palms; Winterhaven; Yermo
F		X	Eureka Valley
G		X	Antioch
H		X	Carson City; Fallon; Honey Lake
I	X		San Quintín

TABLE 6.9.5.
Chromosome Pairing and Stainable Pollen of Intra- and Intertaxon Hybrids

Hybrids	Occurrence of Different Meiotic Configurations		Percent Stainable Pollen		
	7II	r4 + 5II	No. Plants Examined	Range	Mean
C × C	3	—	4	97–98	98
D × D	3	1	22	45–99	92
E × E	3	3[a]	8	55–95	77
A × D	1	1	2	68–76	72
A × E	1	—	1	86	86
B × C	—	8	16	62–96	83
C × F	5	3	14	54–99	93
D × E	6	4	20	32–98	73
D × G	10	2	27	77–99	97
E × G	8	5	33	37–98	78
E × H	1	5	6	67–74	70
			13	28–49	39
G × H	1	—	1	47	47

[a]Obtained from three crosses with a Bermuda Dunes plant as one parent in each combination.

TABLE 6.9.6.
Chromosome Pairing and Stainable Pollen of Intertaxon Hybrids

Hybrids	Occurrence of Different Meiotic Configurations						Percent Stainable Pollen		
	1r4	2r4	1r6	3r4	1r4 + 1r6	1r8	No. Plants Examined	Range	Mean
B × D	5	—	8	—	—	—	18	32–60	47
B × G	1	—	4	—	—	2	11	24–52	42
C × E	3	2	—	—	—	—	2	58–65	62
C × G	—	—	—	—	—	3	6	22–54	37
C × H	—	—	—	2[a]	—	—	1	74	74
							5	34–40	36
F × D	—	—	1	—	1	—	4	42–67	53
F × G	—	1	2	—	—	—	6	40–66	49
I × G	—	—	—	—	—	2	2	52–72	62

[a]Supernumerary chromosome.

Case Study Exercise 10 (Compositae)

PROBLEM: Provide a classification for the seven taxa (A–G), using any or all of the following categories: species, subspecies, variety, form.

MATERIALS AND METHODS: Plants were grown from seeds taken from field collections at the Botanical Gardens of the University of Michigan, Ann Arbor, from 1966 to 1969, and at Davis, California, in 1969 and 1970. In addition, herbarium specimens were examined from CAS, GH, JEPS, ND, UC, and US. Vouchers were deposited at MICH and DAV.

Seeds were germinated by placing them on moistened filter paper in petri dishes kept in the dark at 12–15°C.

In the course of the study, 85 different populations were examined cytologically. PMCs were fixed in fresh solutions of 3:1 (absolute ethanol : glacial acetic acid), stored in a freezer, and studied by the conventional acetocarmine squash technique. Root tips, examined after treatment with 0.1% colchicine for 3–4 hours at 4°C, were similarly fixed, softened by heating 8–12 minutes at 60°C in 1N HCl, and studied by the squash technique with aceto-orcein stain. Estimates of fertility were obtained by examining the stainability of a minimum of 500 pollen grains on one or generally two heads per plant, collected several weeks apart. Grains were placed in acetocarmine and considered viable if they took up stain and were fully expanded.

Experimental hybridizations were made between 28 different populations (table 6.10.1). For each cross 12 to 14 heads were reciprocally cross-pollinated and then bagged. Each population was represented by four to eight plants. Of the 81 hybrid progenies, 31 were analyzed cytogenetically.

Breeding systems were determined by studying pollen tube growth on experimentally selfed and outcrossed florets. The florets were pollinated and then bagged with cotton gauze for six hours. The styles were then removed, heated for one hour at 55°C in a solution of lactophenol with 1% fast green and 1% acid fuchsin, and examined microscopically for the presence of pollen tubes. No pollen tubes were observed in selfed styles whereas dozens of tubes were seen in outcrossed styles. The physiological barrier which prevents self-pollination functions in the stigma, and the pollen is trinucleate. Consequently, these taxa are considered to possess a sporophytic self-incompatibility system (Brewbaker. 1957. *J. Hered.* 48: 271–277).

Morphological studies are based on both field-collected plants and plants grown in a uniform garden in Ann Arbor in 1968. Unfortunately, a high proportion of the plants was killed by cool, moist summer weather; however, three garden populations survived. Measurements of these plants and of plants from field collections demonstrated that the characters studied have a high genetic component, since there were no significant differences between the qualitative characters of plants grown in the garden or in the field (Gottlieb. 1969. Ph.D. Diss. Ann Arbor: Univ. Michigan). Only a single significant difference was noted in the means of four quantitative characters measured in the three sets of populations. The loss of most of the other uniform garden plants made it impossible to treat statistically the quantitative morphological variability. Consequently, although the morphological characters reported herein are sufficient to differentiate the taxa, they provide little information about the substantial amount of variability within each.

Distribution and habitat summary: In general the ranges of these annual taxa are allopatric. But because they occupy habitats which are often juxtaposed, and because several of them colonize roadsides, populations of two or more of the taxa are occasionally encountered in sympatry. This situation has most likely been exacerbated by the extensive road-building activities in the region.

Taxon A extends from the Mohave Desert in California across the southwestern states into western Colorado and New Mexico (figure 6.10.3). It is generally found in sandy soils in sagebrush, creosote bush, and *Coleogyne* shrub communities. Taxon B is also widespread and is encountered in light sandy soils in a diverse range of climatic conditions: equable maritime sites in the Channel Islands and along the California coast near Santa Barbara; arid localities in the inner South Coast Ranges; vineyards in the hot and dry Central Valley south of Fresno; moist shaded forest up to 8,500 ft in the southwestern Sierra Nevada and San Gabriel Mountains; open volcanic soils in the eastern Sierra Nevada up to 9,000 ft; and sandy localities in the cold sagebrush desert of the Great Basin (figure 6.10.3). The two taxa are sometimes neighboringly sympatric in the southern Sierra and in the mountains of southwestern Nevada, with Taxon B found from 5,000 to 9,000 ft and A at lower elevations. The possibility of hybridization between them in these localities is being studied.

Taxon F occupies abandoned sandy fields and chaparral from the vicinity of Los Angeles south to northern Baja California (figure 6.10.3). Taxon E is restricted to south-central California, where it is found in several habitats, including: wind- and salt-swept sand dunes near the ocean; compacted dunes and marine shales further inland; and sand, shale, and serpentine soils in inland sites as high as 3,500 ft (figure 6.10.3).

Taxon C is common in coastal southern California (6.10.4). It is abundant in chaparral openings and dry sandy hills from sea level to 6,000 ft, and has evolved vigorous populations on the gravelly, steep-sided roadcuts of the numerous freeways in the area. Taxon D is typically encountered in chaparral openings north of San Francisco, but where it grows in the arid inner South Coast Ranges, it is generally found in oak savannah and sandy grasslands (figure 6.10.4). In the southern part of its range, it is usually at higher elevations to 6,000 ft. Both C and D are occasionally found on azonal sandstone and serpentine soils.

Taxon G occupies more coastal sites and elevations below 2,000 ft (figure 6.10.4). It is found in pioneer habitats such as old clearings, the landward side of coastal sand dunes, chaparral openings, and sandy roadside embankments, particularly in San Diego County.

DATA: figures 6.10.1 through 6.10.4; tables 6.10.1 through 6.10.6.

REFERENCE: Gottlieb, L. D. 1971. *Evolution* 25: 312–329.

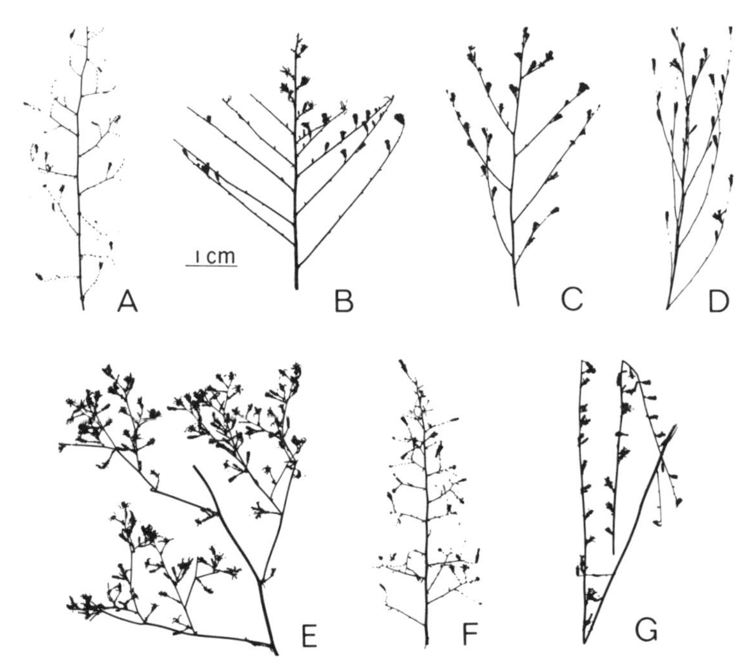

FIGURE 6.10.1 Silhouettes of the branching habits of the seven taxa A–G. All same scale.

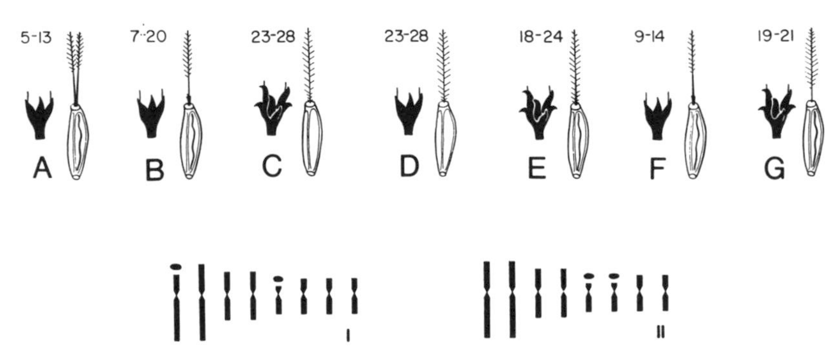

FIGURE 6.10.2 Karyotypes and morphological characteristics of achenes, pappus bristles (including number per achene), and involucral bracts of taxa A–G. Karyotype I is found in taxa A, B, E, and F; karyotype II is found in taxa C, D, and G.

FIGURE 6.10.3 Distributions of taxa A, B, E, and F. Circles, Taxon A; triangles, Taxon B; diamonds, Taxon E; stars, Taxon F.

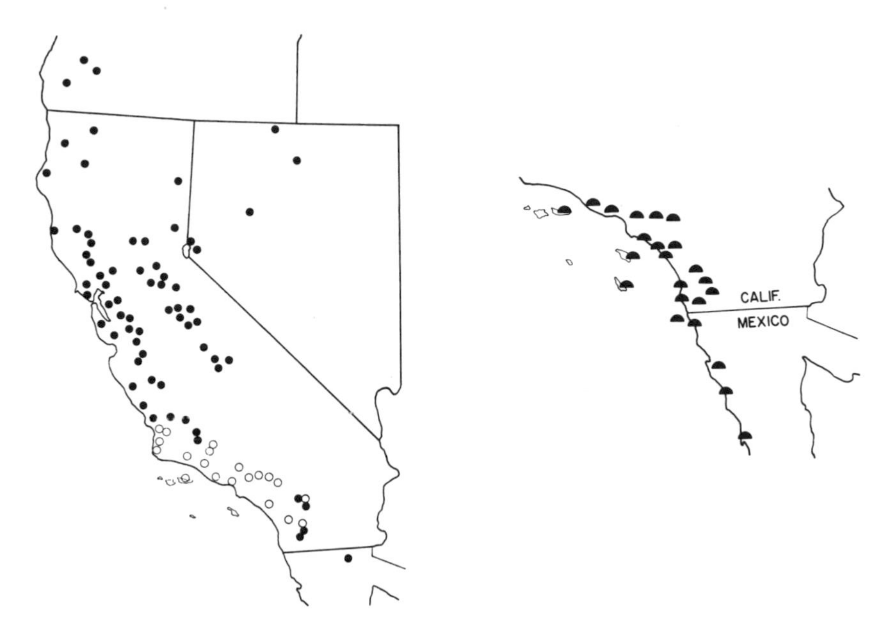

FIGURE 6.10.4 Distributions of taxa C, D, and G. Open circles, Taxon C; blackened circles, Taxon D; blackened hemicircles, Taxon G.

TABLE 6.10.1.
Localities of Populations of Taxa A–G Used as Parents in Experimental Hybridizations

Taxon	Voucher	Locality
A	6650	San Bernardino Co., 21 miles east of Lucerne Valley on Old Woman Springs Rd.
	6655	Riverside Co., Lost Horse Valley, Joshua Tree Nat'l. Mon.
	684	Dona Ana Co., N.M., 1 mile west of Organ on U.S. 70
B	6667	Inyo Co., 13.5 miles west of Bishop on Sabrina Lake Rd.
	6695	Harney Co., Ore., 3 miles south of Narrows on Hwy. 205
	66125	Monterey Co., 0.5 mile west of Salinas River Br. at Gonzales
	66128	Monterey Co., junction of U.S. 101 and Jolon Rd., north of King City
	66153	Ventura Co., on Ozena-Lockwood Rd., 3 miles west of Wagon Rd. #1 Campground
	6843	San Luis Obispo Co., 2 miles east of Pozo Rd. junction on Rt. 58
C	6646	Los Angeles Co., east of Woodfall Rd. near Solemint
	66148	Santa Barbara Co., 0.5 miles west of La Cumbre Lookout
	66154	Ventura Co., Ozena-Lockwood Rd., west of Fraser Park
	66155	Los Angeles Co., Mint Canyon Rd., 7 miles north of Solemint
	66157	Los Angeles Co., 1 mile south of Crystal Lake on Hwy. 39
	66159	San Bernardino Co., Lytle Creek Rd., 0.5 mile south of Applewhite junction
D	6699	Lake Co., Bartlett Springs Rd., north of Lucerne
	66107	Contra Costa Co., near summit of Mt. Diablo
	66108	Contra Costa Co., North Gate junction, Mt. Diablo Rd.
E	66179	San Luis Obispo Co., San Bernardo Creek (from type locality)
	6852	San Luis Obispo Co., south of bridge over San Luis Creek, Avila Beach exit on U.S. 101
	66127	Monterey Co., 1 mile south of Jolon on Jolon-Lockwood Rd.
	6839	Fresno Co., 2 miles east of summit on Hwy. 198
	6845	Kern Co., 0.5 mile east of Temblor Mt. summit on Rt. 58
F	6639	San Diego Co., 1 mile north of Warner Springs on Hwy. 79
	66160	San Diego Co., junction of Hwys. 76 and 56 at Rincon Springs
G	6637	San Diego Co., 4 miles north of Lakeside on Hwy. 67
	66168	San Diego Co., Torrey Pines State Park entrance
	66169	San Diego Co., 19 miles south of San Clemente on I5

NOTE: Collection numbers refer to the author's collections; all counties are in California unless otherwise noted.

TABLE 6.10.2.

Summary of the Morphological Similarities and Differences of Taxa A–G

	Pappus Bristles							Achene Grooves		Bracts		Florets	Branchlet Length[a]			Pubescence[b]	
	Bases																
Taxa	Connate	Separate	Thickened	Intermediate	Not Thickened	Mean No. per Achene	Mean Percent Plumose	Present	Absent	Appressed	Reflexed	Mean no. per head	Long > 15mm	Int. 8–15 mm	Short ≤ 7mm	Glandular	Nonglandular
A	+		+			5–13	43–52	+		+		5–8	+			+	
B	+		+			7–20	61–82	+		+		7–11			+		+
C		+			+	23–28	100		+		+	8–9			+		+
D		+			+	23–28	100		+	+		5–6			+		+
E	+	+	+	+		18–24	88–100	+			+	7–9	+	+			+
F	+		+			9–14	55–61	+		+		7–9	+			+	
G		+			+	19–21	81–84	+			+	11–13			+	+	

[a] Branchlet length refers to the length of the longest primary branchlet on the middle third of a branch emerging approximately halfway up the stem.
[b] Pubescence was scored on involucres and branchlets.

TABLE 6.10.3.
Summary of Pollen Fertilities of F_1 Hybrid Progenies Within and Between Taxa A–G

Cross (Accession Numbers)	No. of F_1 Individuals	Range of Pollen Stainability (%)	Mean Pollen Stainability (%)
C × C			
66148 × 66155	9	89–100	96
66148 × 66157	7	67–98	89
D × D			
6699 × 66107	7	85–100	96
C × D			
6699 × 6646	5	80–90	85
6699 × 66148	2	89–90	90
reciprocal	3	83–84	84
6699 × 66154	4	81–94	89
6699 × 66159	5	77–88	83
66148 × 66107	6	83–91	88
66157 × 6699	1	—	78
66157 × 66107	7	67–85	78
A × A			
684 × 6650	9	77–83	80
684 × 6655	11	54–80	73
B × B			
6667 × 6695	2	34–40	37
6667 × 66153	2	74–84	80
66125 × 6667	6	32–64	46
66125 × 66128	8	85–98	92
66125 × 66153	8	63–83	73
66128 × 6667	2	56–72	64
F × F			
6639 × 66160	7	—	100
E × E			
66127 × 6839	6	90–98	95
6839 × 6852	10	78–96	93
6845 × 6852	10	76–88	82
6852 × 66127	2	95–96	96
A × B			
6650 × 6667	4	41–64	54
6655 × 6667	4	60–73	67
6655 × 66128	1	—	65
6695 × 6655	9	25–63	42
66125 × 6655	6	47–65	59
66153 × 6650	6	69–86	78
reciprocal	9	46–84	70

Cross (Accession Numbers)	No. of F_1 Individuals	Range of Pollen Stainability (%)	Mean Pollen Stainability (%)
A × F			
6639 × 6655	2	70–78	74
6655 × 66160	5	25–50	41
B × F			
6667 × 6639	3	30–41	36
6667 × 66160	1	—	36
66125 × 6639	4	19–64	38
66125 × 66160	3	19–31	25
reciprocal	1	—	29
66153 × 66160	1	—	21
E × A			
6655 × 66127	7	27–44	36
6655 × 6839	5	29–50	39
66179 × 6655	4	35–44	38
E × B			
66128 × 66127	17	58–82	75
6843 × 6845	3	74–85	79
reciprocal	4	73–88	81
66179 × 6667	4	11–24	16
66179 × 66153	6	34–66	48
66125 × 66179	6	38–73	63
E × F			
66179 × 66160	4	27–36	30
A × C			
6655 × 66155	5	8–19	13
6655 × 66157	1	—	8
B × C			
6667 × 66148	6	7–25	16
6667 × 66155	1	—	17
6667 × 66157	1	—	12
66125 × 66148	2	16–20	18
66125 × 66155	1	—	12
66148 × 66153	2	10–19	15
F × C			
66148 × 66160	4	3–13	9
E × C			
6839 × 66157	3	18–26	21
66179 × 66148	8	6–19	13

TABLE 6.10.3 (continued)

Cross (Accession Numbers)	No. of F_1 Individuals	Range of Pollen Stainability (%)	Mean Pollen Stainability (%)	Cross (Accession Numbers)	No. of F_1 Individuals	Range of Pollen Stainability (%)	Mean Pollen Stainability (%)
A × D				**D × G**			
6655 × 66107	3	4–9	7	6699 × 66168	3	0–7	3
B × D				6699 × 66169	2	1–2	2
6667 × 6699	2	4–6	5	reciprocal	1	—	2
6667 × 66107	2	3–7	5	6699 × 6637	2	2	2
6699 × 66128	1	—	6	66169 × 66108	1	—	3
6699 × 66153	1	—	2	**A × G**			
66108 × 66125	1	—	10	6655 × 66168	6	2–11	6
F × D				reciprocal	2	5–8	7
6639 × 66107	3	0–2	1	**B × G**			
E × D				6667 × 66168	3	0–4	2
66127 × 66107	7	6–16	11	66125 × 6637	5	0–5	2
66179 × 6699	3	5–11	8	**F × G**			
C × G				66160 × 66168	2	8–10	9
66148 × 66168	2	1–3	2				
66157 × 66168	1	—	2				

TABLE 6.10.4.
Summary of Chromosome Associations in F_1 Hybrids

Cross	No. Cells Examined	Chromosome Associations (in %)							No. Cells Bridge + Fragment	Mean No. I's/ Cell
		8II	7II+2I	6II+4I	6II+1IV	5II+ 1IV+2I	5II+ 1VI	4II+ 2IV		
C × D										
6699 × 66159	102	90	1		9				0	<0.1
66148 × 6699	112	96	4						0	0.1
66157 × 6699	78	100							0	0.0
A × B										
6650 × 6667	101	100							0	0.0
6655 × 66128	71	85			15				—[a]	0.0
66125 × 6655	98	92	3		5				—[a]	<0.1
A × F										
6655 × 66160	132	80	10		10				5	0.2
B × F										
66125 × 6639	100	85	8		7				2	0.2
66125 × 66160	110	86	1		13				2	<0.1
E × A										
6655 × 6839	67	75	18	3	4				0	0.5
66179 × 6655	62	70	18		12				—[a]	0.4
E × B										
6845 × 6843	86	99	1						0	<0.1
66128 × 66127	71	100							2	0.0
66125 × 66179	146	69	10		16	3	1	1	6	0.3
66179 × 6667	120	68	13		15	2		2	—[a]	0.3
66179 × 66153	87	100							0	0.0
E × F										
66179 × 66160	107	65	22	4	7	2			3	0.6

NOTE: See tables 6.10.5 and 6.10.6 for additional crosses.
[a]Not studied.

TABLE 6.10.5.

Summary of Chromosome Associations in F_1 Hybrids

Cross	No. Cells Examined	Chromosome Associations (in %)										No. Cells Bridge + Fragment	Mean No. I's / Cell
		8II	7II+2I	6II+4I	5II+6I	<5II+ >6I	6II+ 1IV	5II+ 1IV+ 2I	<5II+ 1IV+ >2I	5II+ 1VI	4II+ 2IV		
A × C													
6655 × 66155	47	66	4				30					—[a]	0.1
6655 × 66157	108	42	10				47			1		8	0.2
B × C													
66125 × 66148	73	48	1				48	3				19	0.1
E × C													
6839 × 66157	61	34	41	20			1	3				2	1.7
66179 × 66148	87	46	25	7	5		16	1				—[a]	1.1
A × D													
6655 × 66107	146	56	14	2			22	3	2		1	—[a]	0.5
B × D													
6667 × 6699	120	57	27	7			8			1		9	0.8
6699 × 66128	106	8	13	30	16	21	3	1	7			12	4.5
E × D													
66179 × 6699	101	21	19	32	16	5	3		4			11	3.2

[a] Not studied.

TABLE 6.10.6.

Summary of Chromosome Associations in F_1 Hybrids

Cross	No. Cells Examined	Chromosome Associations (in %)												No. Cells Bridge + Fragment	Mean No. I's/Cell
		8II	7II+2I	6II+4I	5II+6I	<5II+ >6I	6II+ 1IV	5II+ 1IV+ 2I	<5II+ 1IV+ >2I	5II+ 1VI	4II+ 2IV	3II+ 2IV+ 2I	3II+ 1V+1IV +1I		
G × A															
6655 × 66168	78	29	30	22	9	5	1	2						1	2.3
G × B															
6667 × 66168	86	7	16	37	19	20			1					3	4.7
G × F															
66160 × 66168	63	5	25	29	27	9		3	1					1	4.1
G × C															
66148 × 66168	92	15	12	17	14	6	11	10	5	1	1	6	1	10	3.1
G × D															
6699 × 66168	87	1	6	11	30	46	1		5					2	6.7

Literature Cited

Alston, R. A. and B. L. Turner. 1963. *Biochemical Systematics*. Englewood Cliffs, N. J.: Prentice-Hall.

Ashlock, P. D. 1991. Weighting and anagenetic analysis. In E. Mayr and P. D. Ashlock, *Principles of Systematic Zoology,* ed. 2, pp. 236–241. New York: McGraw-Hill.

Babcock, E. B., G. L. Stebbins, Jr., and J. A. Jenkins. 1937. Chromosomes and phylogeny in some genera of the Crepidinae. *Cytologia,* Fujii Jubilee vol. 188–210.

Bauhin, G. 1623. *Pinax Theatri Botanici.* Basel.

Behnke, H.-D. 1986. Sieve-element characters and the systematic position of *Austrobaileya,* Austrobaileyaceae—with comments to the distinction and definition of sieve cells and sieve-tube members. *Pl. Syst. Evol.* 152: 101–121.

Berlin, B. 1992. *Ethnobiological Classification: Principles of Categorization of Plants and Animals in Traditional Societies.* Princeton: Princeton University Press.

Brown, G. K. 1983. Chromosome numbers in *Platyschkuhria* Rydberg (Compositae) and their systematic significance. *Amer. J. Bot.* 70: 591–601.

Buck, R. C. and D. L. Hull. 1966. The logical structure of the Linnaean hierarchy. *Syst. Zool.* 15: 97–111.

Burma, B. H. 1954. Reality, existence, and classification: A discussion of the species problem. *Madroño* 12: 193–209.

Candolle, A. L. P. P. de, ed. 1844–1873. *Prodromus Systematis Naturalis Regni Vegetabilis,* vols. 8–17. Paris.

Candolle, A. P. de. 1813. *Theorie Elementaire de la Botanique*. Paris.

Candolle, A. P. de, ed. 1824–1838. *Prodromus Systematis Naturalis Regni Vegetabilis,* vols. 1–7. Paris.

Chater, A. O. and R. K. Brummitt. 1966a. Subspecies in the works of Friedrich Ehrhart. *Taxon* 15: 95–106.

——. 1966b. Subspecies in the works of Christiaan Hendrick Persoon. *Taxon* 15: 143–149.

Claugher, D., ed. 1990. *Scanning Electron Microscopy in Taxonomy and Functional Morphology.* Oxford: Clarendon Press.

Clausen, J., D. D. Keck, and W. M. Hiesey. 1940. Experimental studies on the nature of species I: Effect of varied environments on western North American plants. *Publ. Carnegie Inst.* (Washington) 520: i–viii, 1–452.

Clausen, R. T. 1941. On the use of the terms "subspecies" and "variety." *Rhodora* 43: 157–167.

Crawford, D. J. 1990. *Plant Molecular Systematics: Macromolecular Approaches.* New York: Wiley.

Crawford, D. J. and H. D. Wilson. 1979. Allozyme variation in several closely related diploid species of *Chenopodium* of the western United States. *Amer. J. Bot.* 66: 237–244.

Cronquist, A. 1981. *An Integrated System of Classification of Flowering Plants.* New York: Columbia University Press.

——. 1988. *The Evolution and Classification of Flowering Plants.* 2d ed. New York: New York Botanical Garden.

Crovello, T. J. 1970. Analysis of character variation in ecology and systematics. *Ann. Rev. Ecol. Syst.* 1: 55–98.

——. 1974. Analysis of character variation in systematics. In A. E. Radford, W. C. Dickison, J. R. Massey, and C. R. Bell, eds., *Vascular Plant Systematics,* pp. 451–484. New York: Harper and Row.

Darwin, C. 1859. *On the Origin of Species by Means of Natural Selection.* London: Murray.

Donoghue, M. J. 1985. A critique of the biological species concept and recommendations for a phylogenetic alternative. *Bryologist* 88: 172–181.

Doyen, J. T. and C. N. Slobodchikoff. 1974. An operational approach to species classification. *Syst. Zool.* 23: 239–247.

Doyle, J. J. 1993. DNA, phylogeny, and the flowering of plant systematics. *Bioscience* 43: 380–389.

Eggers Ware, D. M. 1983. Genetic fruit polymorphism in North American *Valerianella* (Valerianaceae) and its taxonomic implications. *Syst. Bot.* 8: 33–44.

Ehrlich, P. R. and P. H. Raven. 1969. Differentiation of populations. *Science* 165: 1228–1232.

Ereshefsky, M., ed. 1992. *The Units of Evolution: Essays on the Nature of Species.* Cambridge: MIT Press.

Estabrook, G. F. 1986. Evolutionary classification using convex phenetics. *Syst. Zool.* 35: 560–570.

Goodspeed, T. H. 1934. *Nicotiana* phylesis in the light of chromosome number, morphology, and behavior. *Univ. California Publ. Bot.* 17: 369–398.

Gottlieb, L. D. 1981. Electrophoretic evidence and plant populations. *Progr. Phytochem.* 7: 1–46.

Grant, V. 1981. *Plant Speciation*. 2d ed. New York: Columbia University Press.
Gray, A. 1836. *Elements of Botany*. New York: Carvill.
Greuter, W. et al., eds. 1988. International Code of Botanical Nomenclature as Adopted by the Fourteenth International Botanical Congress, Berlin, July-August 1987. *Regnum Veg*. 118: i–xiv, 1–328.
Grew, N. 1682. *The Anatomy of Plants*. London.
Hall, A. V. 1991. A unifying theory for methods of systematic analysis. *Biol. J. Linn. Soc*. 42: 425–456.
Hall, H. M. 1929. The taxonomic treatment of units smaller than species. *Proc. Internl. Congr. Pl. Sci*. 2: 1461–1468.
Hamilton, C. W. and S. H. Reichard. 1992. Current practice in the use of subspecies, variety, and forma in the classification of wild plants. *Taxon* 41: 485–498.
Hennig, W. 1966. *Phylogenetic Systematics*. Translated by D. D. Davis and R. Zangerl. Urbana: University of Illinois Press.
Hopwood, A. T. 1959. The development of pre-Linnaean taxonomy. *Proc. Linn. Soc. (London)* 170: 230–234.
Jussieu, A. L. de. 1789. *Genera Plantarum Secundum Ordines Naturales Disposita*. Paris.
Kapadia, Z. J. 1963. Varietas and subspecies: A suggestion towards greater uniformity. *Taxon* 12: 257–259.
Knobloch, I. W. 1972. Intergeneric hybridization in flowering plants. *Taxon* 21: 97–103.
Levin, D. A. 1979. The nature of plant species. *Science* 204: 381–384.
——. 1993. Local speciation in plants: The rule not the exception. *Syst. Bot*. 18: 197–208.
Lindley, J. 1830–1840. *The Genera and Species of Orchidaceous Plants*. London: Ridgways.
Linnaeus, C. 1735. *Systema Naturae*. Leiden.
——. 1737. *Genera Plantarum*. Leiden.
——. 1751. *Philosophia Botanica*. Stockholm.
——. 1753. *Species Plantarum*. 2 vols. Stockholm.
Mayr, E. 1940. Speciation phenomena in birds. *Amer. Nat*. 74: 249–278.
——. 1963. *Animal Species and Evolution*. Cambridge: Harvard University Press.
——. 1968. The role of systematics in biology. *Science* 159: 595–599.
Ornduff, R. 1967. *Papers on Plant Systematics*. Boston: Little, Brown.
Palmer, J. D. 1987. Chloroplast DNA evolution and biosystematic uses of chloroplast DNA variation. *Amer. Nat*. 130 (suppl.): s6–s29.
Persoon, C. H. 1805. *Synopsis Plantarum, seu Enchiridium Botanicum*, vol. 1. Paris.
Ramsbottom, J. 1938. Linnaeus and the species concept. *Proc. Linn. Soc. (London)* 150: 192–219.
Robichaux, R. H. and J. E. Canfield. 1985. Tissue elastic properties of eight Hawaiian *Dubautia* species that differ in habitat and diploid chromosome number. *Oecologia* 66: 77–80.
Rollins, R. C. 1953. Cytogenetical approaches to the study of genera. *Chron. Bot*. 14: 133–139.

Sanderson, M. J. and J. J. Doyle. 1993. Phylogenetic relationships in North American *Astragalus* (Fabaceae) based on chloroplast DNA restriction site variation. *Syst. Bot.* 18: 395–408.

Saville, D. B. O. 1975. Evolution and biogeography of Saxifragaceae and guidance from their rust parasites. *Ann. Missouri Bot. Gard.* 62: 354–361.

Schnarf, K. 1931. *Vergleichende Embryologie der Angiospermen.* Berlin: Borntraeger.

Sheviak, C. J. and M. L. Bowles. 1986. The prairie fringed orchids: A pollinator–isolated species pair. *Rhodora* 88: 267–290.

Simpson, G. G. 1961. *Principles of Animal Taxonomy.* New York: Columbia University Press.

Sinclair, W. A. 1951. *The Traditional Formal Logic: A Short Account for Students.* 5th ed. London: Methuen.

Sneath, P. H. A. and R. R. Sokal. 1973. *Numerical Taxonomy: The Principles and Practice of Numerical Classification.* San Francisco: Freeman.

Sokal, R. R. 1965. Statistical methods in systematics. *Biol. Rev. Cambridge Philos. Soc.* 40: 337–391.

Sokal, R. R. and T. J. Crovello. 1970. The biological species concept: A critical evaluation. *Amer. Nat.* 104: 127–153.

Sokal, R. R. and F. J. Rohlf. 1981. *Biometry: The Principles and Practice of Statistics in Biological Research.* 2d ed. San Francisco: Freeman.

Stebbins, G. L. 1977. *Processes of Organic Evolution.* 3d ed. Englewood Cliffs, New Jersey.: Prentice-Hall.

Stuessy, T. F. 1977. Case studies: A new approach to teaching plant systematics. *Sida* 7: 107–117.

——. 1979. Ultrastructural data for the practicing plant systematist. *Amer. Zoologist* 19: 621–636.

——. 1987. Explicit approaches for evolutionary classification. *Syst. Bot.* 12: 251–162.

——. 1989. Comments on specific categories in flowering plants. *Pl. Syst. Evol.* 167: 69–74.

——. 1990. *Plant Taxonomy: The Systematic Evaluation of Comparative Data.* New York: Columbia University Press.

Thorne, R. F. 1976. A phylogenetic classification of the Angiospermae. *Evol. Biol.* 9: 35–106.

——. 1992. Classification and geography of the flowering plants. *Bot. Review* 58: 225–348.

Tournefort, J. P. de. 1700. *Institutiones Rei Herbariae, editio altera.* Paris.

Tufte, E. R. 1983. *The Visual Display of Quantitative Information.* Cheshire, Conn.: Graphics Press.

Valentine, D. H. 1975. The taxonomic treatment of polymorphic variation. *Watsonia* 10: 385–390.

Warburton, F. E. 1967. The purposes of classifications. *Syst. Zool.* 16: 241–245.

Index

Numbers in **boldface** refer to tables or figures. Numbers in [brackets] refer to entries represented by "et al." in the text.